Fuel Options for Reducing Greenhouse Gas Emissions from Motor Vehicles

Final Report September 2003

U.S. Department of Transportation

Notice

This document is disseminated under the sponsorship of the Department of Transportation in the interest of information exchange. The United States Government assumes no liability for its contents or use thereof.

REPORT DOCUMENTATION PAGE			Form Approved OMB No. 0704-0188
Public reporting burden for this collection of information is estimated to average 1 hour per response, including the time for reviewing instructions, searching existing data sources, gathering and maintaining the data needed, and completing and reviewing the collection of information. Send comments regarding this burden estimate or any other aspect of this collection of information, including suggestions for reducing this burden, to Washington Headquarters Services, Directorate for Information Operations and Reports, 1215 Jefferson Davis Highway, Suite 1204, Arlington, VA 22202-4302, and to the Office of Management and Budget, Paperwork Reduction Project (0704-0188), Washington, DC 20503			
1. AGENCY USE ONLY (Leave blank)	2. REPORT DATE September 2003	3. REPORT TYPE AND DATES COVERED Final Report October 2001-August 2003	
4. TITLE AND SUBTITLE Fuel Options for Reducing Greenhouse Gas Emissions from Motor Vehicles			P3025/RS391
6. AUTHOR(S) Don Pickrell		5. FUNDING NUMBERS	
7. PERFORMING ORGANIZATION NAME(S) AND ADDRESS(ES) U.S. Department of Transportation John A. Volpe National Transportation Systems Center 55 Broadway Cambridge, MA 02142		8. PERFORMING ORGANIZATION REPORT NUMBER DOT-VNTSC-RSPA-03-03	
9. SPONSORING/MONITORING AGENCY NAME(S) AND ADDRESS(ES) U.S. Department of Transportation Research and Special Programs Administration 400 7th Street, S.W. Washington, D.C. 20590		10. SPONSORING/MONITORING AGENCY REPORT NUMBER	
11. SUPPLEMENTARY NOTES			
12a. DISTRIBUTION/AVAILABILITY STATEMENT This document is available to the public through the National Technical Information Service, Springfield, Virginia 22161.		12b. DISTRIBUTION CODE	
13. ABSTRACT (Maximum 200 words) This report assesses the potential of substitutes for gasoline to reduce emissions of carbon dioxide and other greenhouse gases (GHGs) by automobiles and light-duty trucks. It estimates reductions in future GHG emissions under specific assumptions about growth in light-duty vehicle travel and the replacement of gasoline by various other fuels, both in the near term (10 years) and over the longer term (25 years). Under reasonable assumptions about the fraction of projected gasoline use that could be replaced by another fuel within these time horizons, it concludes that the reduction in GHG emissions from most gasoline substitutes would be modest. The report also assesses the cost-effectiveness of replacing gasoline with each of these fuels as a strategy for reducing GHG emissions, and concludes that promoting alternative fuels would be a costly strategy for reducing emissions. Finally, the study also briefly surveys other concerns that are likely to arise in making a transition from near-exclusive reliance on gasoline to widespread production and use of any alternative fuel. These concerns include potential health and safety consequences, developing the infrastructure required to support commercial-scale production and distribution of gasoline substitutes, and producing adequate supplies of feedstocks required to refine certain fuels.			
14. SUBJECT TERMS Alternative fuels, greenhouse gases, vehicle emissions, transportation fuels		15. NUMBER OF PAGES	
		16. PRICE CODE 76	
17. SECURITY CLASSIFICATION OF REPORT Unclassified	18. SECURITY CLASSIFICATION OF THIS PAGE Unclassified	19. SECURITY CLASSIFICATION OF ABSTRACT Unclassified	20. LIMITATION OF ABSTRACT Unlimited

METRIC/ENGLISH CONVERSION FACTORS

ENGLISH TO METRIC

LENGTH (APPROXIMATE)
- 1 inch (in) = 2.5 centimeters (cm)
- 1 foot (ft) = 30 centimeters (cm)
- 1 yard (yd) = 0.9 meter (m)
- 1 mile (mi) = 1.6 kilometers (km)

AREA (APPROXIMATE)
- 1 square inch (sq in, in^2) = 6.5 square centimeters (cm^2)
- 1 square foot (sq ft, ft^2) = 0.09 square meter (m^2)
- 1 square yard (sq yd, yd^2) = 0.8 square meter (m^2)
- 1 square mile (sq mi, mi^2) = 2.6 square kilometers (km^2)
- 1 acre = 0.4 hectare (he) = 4,000 square meters (m^2)

MASS - WEIGHT (APPROXIMATE)
- 1 ounce (oz) = 28 grams (gm)
- 1 pound (lb) = 0.45 kilogram (kg)
- 1 short ton = 2,000 pounds (lb) = 0.9 tonne (t)

VOLUME (APPROXIMATE)
- 1 teaspoon (tsp) = 5 milliliters (ml)
- 1 tablespoon (tbsp) = 15 milliliters (ml)
- 1 fluid ounce (fl oz) = 30 milliliters (ml)
- 1 cup (c) = 0.24 liter (l)
- 1 pint (pt) = 0.47 liter (l)
- 1 quart (qt) = 0.96 liter (l)
- 1 gallon (gal) = 3.8 liters (l)
- 1 cubic foot (cu ft, ft^3) = 0.03 cubic meter (m^3)
- 1 cubic yard (cu yd, yd^3) = 0.76 cubic meter (m^3)

TEMPERATURE (EXACT)
$[(x-32)(5/9)]\ °F = y\ °C$

METRIC TO ENGLISH

LENGTH (APPROXIMATE)
- 1 millimeter (mm) = 0.04 inch (in)
- 1 centimeter (cm) = 0.4 inch (in)
- 1 meter (m) = 3.3 feet (ft)
- 1 meter (m) = 1.1 yards (yd)
- 1 kilometer (km) = 0.6 mile (mi)

AREA (APPROXIMATE)
- 1 square centimeter (cm^2) = 0.16 square inch (sq in, in^2)
- 1 square meter (m^2) = 1.2 square yards (sq yd, yd^2)
- 1 square kilometer (km^2) = 0.4 square mile (sq mi, mi^2)
- 10,000 square meters (m^2) = 1 hectare (ha) = 2.5 acres

MASS - WEIGHT (APPROXIMATE)
- 1 gram (gm) = 0.036 ounce (oz)
- 1 kilogram (kg) = 2.2 pounds (lb)
- 1 tonne (t) = 1,000 kilograms (kg) = 1.1 short tons

VOLUME (APPROXIMATE)
- 1 milliliter (ml) = 0.03 fluid ounce (fl oz)
- 1 liter (l) = 2.1 pints (pt)
- 1 liter (l) = 1.06 quarts (qt)
- 1 liter (l) = 0.26 gallon (gal)
- 1 cubic meter (m^3) = 36 cubic feet (cu ft, ft^3)
- 1 cubic meter (m^3) = 1.3 cubic yards (cu yd, yd^3)

TEMPERATURE (EXACT)
$[(9/5)y + 32]\ °C = x\ °F$

QUICK INCH - CENTIMETER LENGTH CONVERSION

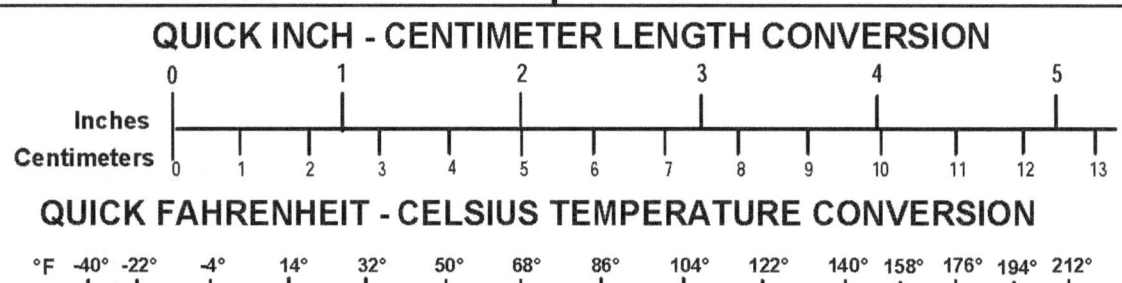

QUICK FAHRENHEIT - CELSIUS TEMPERATURE CONVERSION

°F	-40°	-22°	-4°	14°	32°	50°	68°	86°	104°	122°	140°	158°	176°	194°	212°
°C	-40°	-30°	-20°	-10°	0°	10°	20°	30°	40°	50°	60°	70°	80°	90°	100°

For more exact and or other conversion factors, see NIST Miscellaneous Publication 286, Units of Weights and Measures. Price $2.50 SD Catalog No. C13 10286

TABLE OF CONTENTS

Section	Page
Executive Summary	vii
Energy Use and Greenhouse Gas Emissions	1
What Are Alternative Fuels?	2
Why the Interest in Alternative Fuels?	2
How Alternative Fuels Can Reduce GHG Emissions	2
Scope of this Report	3
Critical Assumptions	4
Growth in Motor Vehicle Use	4
Fuel Use and Engine Technology	5
Greenhouse Gas Reductions from Light-Duty Vehicles	7
How the Estimates Were Developed	7
Near-Term Energy Use and GHG Emissions	8
Longer-Term GHG Emissions Reductions from Alternative Fuels	14
Cost-Effectiveness of Alternative Fuels	24
Major Elements of Alternative Fuel Costs	25
Costs for Near-Term Alternative Fuels	26
Alternative Fuel Cost-Effectiveness	29
Improving the Cost-Effectiveness of Alternative Fuels	30
Potential Concerns with Alternative Fuels	32
Fuel and Feedstock Production Demands	32
Fueling Infrastructure Requirements	36
Potential Hazards from Alternative Fuels	38
Conclusions	40
Appendix	45

LIST OF FIGURES

Figure		Page
ES-1.	Estimated 2010 GHG emissions from light-duty vehicles with exclusive gasoline use and 10% alternative fuel replacement	xi
ES-2.	Estimated 2025 GHG emissions from light-duty vehicles with exclusive gasoline use and 25% alternative fuel replacement	xiii

LIST OF TABLES

Table		Page
ES-1.	Summary of key assumptions used in study	ix
ES-2.	Year 2010 greenhouse gas emissions with all-gasoline baseline and 10% gasoline replacement by alternative fuels	x
ES-3.	Year 2025 greenhouse gas emissions with all-gasoline baseline and 25% gasoline replacement by alternative fuels	xiv
ES-4.	Reductions in GHG emissions from alternative fuels and advanced engine technologies (% changes relative to gasoline in conventional ICE)	xv
ES-5.	Cost-effectiveness of near-term alternative fuels in reducing GHG emissions from light-duty vehicles	xvi
1.	Summary of key assumptions used in study	5
2.	Energy efficiency of gasoline baseline and near-term alternative fuels for light-duty vehicles	9
3.	Greenhouse gas emissions rates for light-duty vehicles using gasoline and near-term alternative fuels	11
4.	Year 2010 greenhouse gas emissions with all-gasoline baseline and 10% gasoline replacement	12
5.	Criteria pollutant emission rates for gasoline and near-term alternative fuel/engine technology combination	14
6.	Energy efficiency of gasoline and long-term alternative fuels for light-duty vehicles	16
7.	Greenhouse gas emissions rates for light-duty vehicles using gasoline and long-term alternative fuel/engine technology combinations	18
8.	Year 2025 greenhouse gas emissions with all-gasoline baseline and 25% gasoline replacement by alternative fuels	20
9.	Criteria pollutant emissions for gasoline and long-term alternative fuel/engine technology combinations	21
10.	Energy efficiency and GHG emissions from alternative fuels and advanced engine technologies (% changes relative to gasoline in conventional ICE)	23
11.	Estimated capital costs for alternative fuel vehicle production and infrastructure provision	27
12.	Cost-effectiveness of near-term alternative fuels in reducing GHG emissions from light-duty vehicles	30
13.	Alternative fuel production volumes necessary to replace 10% of projected 2010 gasoline use	33
14.	Feedstock demands for alternative fuel production volumes necessary to replace 10% of projected 2010 gasoline use	35
15.	Alternative fuel volumes necessary to replace 25% of projected 2025 gasoline use	36
16.	Feedstock demands for alternative fuel volumes necessary to replace 25% of projected 2025 gasoline use	37
17.	Properties of gasoline and alternative fuels affecting potential safety and environmental hazards	39

LIST OF TABLES (cont.)

Table		Page
A-1.	Energy consumption rates for LDV classes operating on near-term alternative fuel/engine technology combinations	47
A-2.	Greenhouse gas emissions rates for LDV classes with near-term alternative fuel/engine technology combinations	48
A-3.	Criteria pollutant emission rates for LDV classes operating on near-term alternative fuel/engine technology combinations	49
A-4.	Energy consumption rates for LDV classes operating on long-term alternative fuel/engine technology combinations	50
A-5.	Greenhouse gas emissions rates for LDV classes operating on long-term alternative fuel/engine technology combinations	51
A-6.	Criteria pollutant emission rates for LDV classes operating on long-term alternative fuel/engine technolgoy combinations	52
A-7.	Near-term light-duty vehicle manufacturing, fuel production, and fuel infrastructure costs for alternative fuel use	53
A-8.	Near-term alternative fuel and feedstock production and energy content	54
A-9.	Long-term alternative fuel and feedstock production and energy content	55

Fuel Options for Reducing Greenhouse Gas Emissions from Motor Vehicles

Executive Summary

This report focuses on the potential of various substitutes for gasoline to reduce energy use as well as emissions of carbon dioxide and other greenhouse gases (GHGs) by automobiles and light-duty trucks. It estimates reductions in future GHG emissions under specific assumptions about growth in light-duty vehicle travel and the replacement of gasoline by various other fuels, both in the near term and over a longer-term horizon. The analysis assumes that an alternative fuel could replace 10% of projected gasoline consumption within a decade, and could substitute for as much as one-quarter of forecast gasoline use by the year 2025. While the estimates of GHG emissions reductions from alternative fuels assume that light-duty vehicles will continue to be powered by internal-combustion engines, the report also explores potential emissions reductions from widespread use of advanced-technology vehicles, including those powered by direct-injection internal combustion engines, hybrid electric drive systems, or fuel cells.

This report also examines whether potential substitutes for gasoline would enable light-duty vehicles to comply with pending federal standards for emissions of "criteria" air pollutants. In addition, it estimates costs for producing and distributing various non-gasoline fuels and for producing vehicles capable of using them, and uses these estimates to assess the cost-effectiveness of replacing gasoline with each of these fuels as a strategy for reducing GHG emissions. Finally, the study also briefly surveys other concerns that are likely to arise in making a transition from the nation's current near-exclusive reliance on gasoline to widespread production and use of various alternatives. These concerns include potential health and safety consequences, developing the infrastructure required to support commercial-scale production and distribution of gasoline substitutes, and producing the feedstocks required to refine certain fuels.

This study differs from most previous research on alternatives to gasoline because it considers diesel fuels -- derived entirely from petroleum or partly from biomass (called bio-diesel) -- to be possible substitutes for gasoline. Another important distinction is that while most previous research estimates potential GHG reductions on a per-vehicle basis or from replacing some arbitrarily assumed quantity of gasoline, this study's estimates of GHG emissions reductions are quantified in total (metric) tons during specific future years. Thus they can be readily compared to current or projected future levels of transportation-related and economy-wide GHG emissions, or to possible targets for their reduction.

This report draws several major conclusions:

- When emissions in fuel production and distribution as well as from vehicle operation are accounted for, replacing 10% of projected gasoline use during the year 2010 with ethanol produced from corn, or diesel fuel derived from petroleum or partly from biomass (bio-diesel) could reduce GHG emissions from light-duty vehicle use by 2-3%. Replacing the same fraction of gasoline use with electricity used in battery-powered electric vehicles could produce comparable GHG reductions, but would require rapid advances in battery technology within the current decade.

- Replacing one-quarter of projected gasoline consumption with petroleum diesel, bio-diesel, or electricity, which is assumed to be possible within a 25-year time horizon, could reduce combined GHG emissions from vehicle use and fuel production by about 8-11%. Net emissions reductions more than twice as large could be achieved if the technology for producing ethanol from cellulosic biomass rather than from corn (the current ethanol feedstock) could be commercialized. Emissions reductions from replacing 25% of gasoline with CNG, LPG, or hydrogen would amount to 5-6% when emissions during both vehicle use and fuel production are accounted for.

- Potential increases in emissions of nitrogen oxides and fine particulates are important concerns with substituting petroleum diesel or bio-diesel for gasoline as a light-duty vehicle fuel. Controversy remains about whether light-duty diesel vehicles will be able to comply with the pending Tier 2 federal emission standards for these pollutants, even with the sharp reduction in the sulfur content of petroleum diesel fuel mandated by recently-adopted EPA regulations.

- Innovations in engine technology appear to have as much potential to reduce GHG emissions as do alternative fuels, and operating vehicles using these technologies on alternative fuels could produce very large reductions in emissions. However, mass production and purchases of vehicles incorporating the most advanced technologies remain uncertain prospects even within this study's long-term horizon.

- Costs for replacing a significant fraction of gasoline use with another fuel -- including additional costs for manufacturing vehicles capable of using it, investments in new or expanded facilities to produce and distribute it, and costs for producing the fuel itself -- would be very substantial. When combined with the reductions in GHG emissions estimated in this study, these costs appear likely to make most alternative fuels unattractive strategies for reducing GHG emissions from a cost-effectiveness standpoint.

- A ubiquitous retailing or other distribution infrastructure comparable to that now offered by retail gasoline stations would also be required to allow any alternative fuel to replace a significant fraction of gasoline use. While some alternative fuels could make extensive use of gasoline distribution and fueling infrastructure already in place with relatively minor modifications, other fuels would require major investments to develop entirely new facilities for their distribution, storage, and retailing.

- Gasoline as well as substitute fuels can present safety risks, potential hazards to human health, and possible environmental damages arising from their production, storage, and distribution, from normal vehicle operation and refueling, and from accidental fuel releases. Risk profiles presented by different fuels indicate that conventional fuels such as gasoline are not free from hazards, and also show that most alternative fuels are not inherently more "dangerous" than the fuels they would replace.

Critical Assumptions

This study's results depend partly on a number of critical assumptions used in the analysis, which are summarized in Table ES-1. As the table shows, light-duty vehicle travel is projected to continue growing at its recent moderate pace, while the recent substitution of light-duty trucks – pickups, vans, and sport-utility vehicles – for automobiles is expected to continue throughout this period. Smaller light truck models (those weighing under 6,000 pounds) are expected to comprise about two-thirds of total light trucks in use over most of this period, with the remaining third made up of larger pickups, sport-utility vehicles, and vans.

Internal combustion engines are expected to remain the predominant technology utilized by the U.S. light-duty vehicle fleet regardless of the combination of gasoline and substitute fuels in use. Fleetwide energy efficiency is expected to increase slightly, reflecting continued evolutionary improvements in the energy efficiency of internal combustion engines, as well as a slight increase in the share of smaller light truck models relative to their larger counterparts. Without government policies or other measures to promote alternative fuels, gasoline is assumed to remain the dominant fuel for light-duty vehicles.[1]

Table ES-1. Summary of key assumptions used in study.

Assumption	Date	Light-Duty Vehicles
VMT Growth	2000-2010	2.5% per year
	2010-2025	2.1% per year
VMT Mix	2000	Passenger Cars: 63% Small Light Trucks: 27% Large Light Trucks: 10%
	2010	Passenger Cars: 55% Small Light Trucks: 33% Large Light Trucks: 12%
	2025	Passenger Cars: 47% Small Light Trucks: 39% Large Light Trucks: 14%
Baseline Fuel	2010	Gasoline (70% Conventional, 30% Federal Phase 2 RFG) (1)
	2025	Gasoline (70% Conventional, 30% Federal Phase 2 RFG) (1)
Baseline Engine Technology	2010	Spark-Ignition Internal Combustion
	2025	Spark-Ignition Internal Combustion
Baseline Energy Efficiency Improvement (2)	2000-2010	Vehicle: 1.2% Full Fuel Cycle: 1.2%
	2010-2025	Vehicle: 4.0% Full Fuel Cycle: 3.8%
Alternative Fuel Use (% of VMT) (2)	2010	10%
	2025	25%

(1) Sulfur content reduced in compliance with provisions of EPA "Tier 2" rule on gasoline sulfur content.

(2) Average for all light-duty vehicle travel, consisting of assumed mix of vehicle classes shown above.

The potential substitutes for gasoline considered in this study for use by light-duty vehicles include ethanol (blended with 10-15% gasoline), compressed natural gas, liquid petroleum gas (propane), petroleum-based diesel, a blend of 80% petroleum and 20% soy-based diesel (bio-diesel), electricity used in battery-powered electric vehicles, and over the longer term, hydrogen. Within the year-2010 near-term horizon, the study assumes that public policies or other concerted efforts to promote these fuel alternatives could result in one or a combination of these fuels substituting for 10% of projected gasoline vehicle travel. Over the 25-year horizon, this study assumes that one -- or again, some combination -- of the alternatives to gasoline assumed to be widely available could substitute for 25% of projected gasoline use.

Potential Reductions in GHG Emissions

Only a few of the alternative fuels analyzed in this study appear to offer the potential to reduce total GHG emissions from light-duty vehicles significantly, with the limited replacement of gasoline assumed to be feasible within this study's near-term horizon.[2] Table ES-2 and

Figure ES-1 summarize the potential reductions in GHG emissions for each of the fuels during 2010. Replacing 10% of projected gasoline use during 2010 with corn-based ethanol is estimated to reduce GHG emissions from vehicle operations by 9%, but higher energy use and emissions (compared to gasoline) during ethanol production limit GHG reductions to about 3% when measured over the full fuel cycle. This latter figure adds GHG emissions that occur during feedstock extraction, fuel refining, storage, distribution, and vehicle fueling – sometimes termed "upstream" or "well

Table ES-2. Year 2010 greenhouse gas emissions with all-gasoline baseline and 10% gasoline replacement by alternative fuels.

Fuel Mix	Total GHG Emissions (Tg CO_2 equivalent) (1)		Reduction from Baseline (Tg)		% Reduction from Baseline	
	Vehicle Operation	Full Fuel Cycle	Vehicle Operation	Full Fuel Cycle	Vehicle Operation	Full Fuel Cycle
100% Gasoline	1,373	1,752	--	--	--	--
10% Ethanol (as E85), 90% Gasoline	1,250	1,701	122	51	-9%	-3%
10% CNG (Bi-Fuel Vehicles), 90% Gasoline	1,362	1,743	10	8	-1%	0%
10% CNG (Dedicated Vehicles), 90% Gasoline	1,359	1,739	14	13	-1%	-1%
10% LPG, 90% Gasoline	1,368	1,733	4	19	0%	-1%
10% Diesel, 90% Gasoline	1,345	1,709	28	43	-2%	-2%
10% Bio-Diesel (B20), 90% Gasoline	1,347	1,713	26	39	-2%	-2%
10% Electricity, 90% Gasoline	1,235	1,705	137	47	-10%	-3%

(1) Teragrams of CO_2 equivalent, computed using 100-year GWPs of 21 for CH_4 and 310 for N_2O. One teragram equals 10^{12} grams, or one million metric tons.

to tank" activities – to those during actual vehicle operation, often referred to as "tailpipe" or "tank to wheel" emissions.[3] Reductions in tailpipe GHG emissions for ethanol appear particularly large because carbon sequestration that occurs during growth of its corn feedstock is credited against tailpipe emissions of ethanol-fueled vehicles, rather than against other components of upstream emissions.

Reductions in tailpipe GHG emissions from replacing 10% of gasoline use with petroleum-based diesel would be considerably smaller (about 2%) than those from ethanol, but would still be roughly comparable over the full fuel cycle because energy use and GHG emissions in petroleum extraction and diesel fuel refining are not only lower than those from growing corn and producing ethanol, but also lower than those from gasoline refining. Replacing 10% of

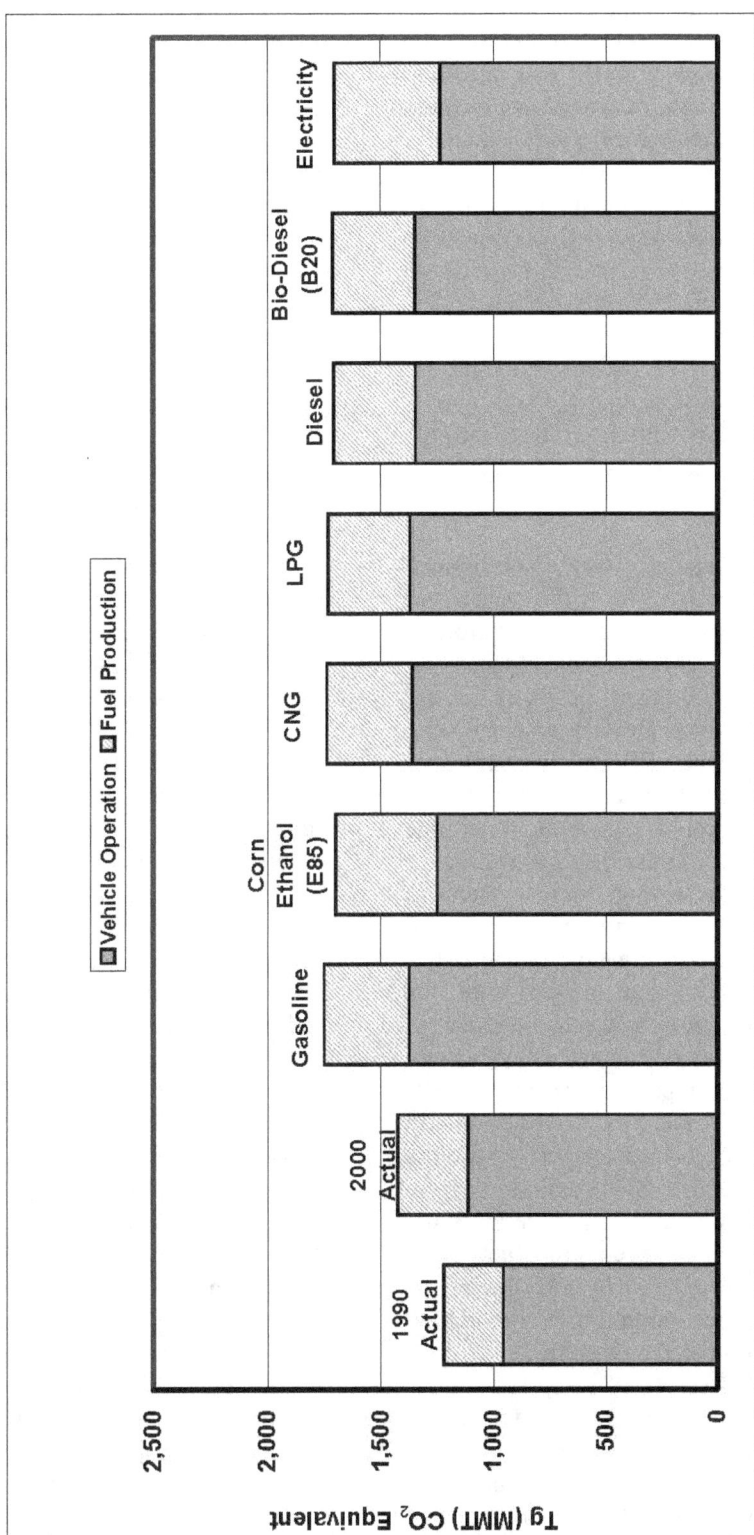

Figure ES-1. Estimated 2010 GHG emissions from light-duty vehicles with exclusive gasoline use and 10% alternative fuel replacement.

gasoline use with bio-diesel (consisting 80% of petroleum diesel and 20% of a soy-based component) would reduce GHG emissions by about 2% in both vehicle operation and over the complete fuel production and use cycle. Since battery-powered electric vehicles do not produce tailpipe emissions, replacing 10% of gasoline-powered light-duty vehicle travel with them would reduce GHG emissions correspondingly, as Table ES-2 shows. However, significant GHG emissions from generating electricity using the mix of fuels expected during 2010 would limit the reduction in emissions to about than 3% when measured over the complete fuel cycle. The remaining fuels included in Table ES-2 and Figure ES-1 – compressed natural gas (CNG) and liquid petroleum gas (LPG) – appear to have very limited potential to reduce tailpipe and fuel cycle GHG emissions (between 0% and 1%).

Significantly larger GHG emissions reductions appear possible with the changes in feedstock sources and processing, evolution of internal-combustion engine technology, and more extensive replacement of gasoline use assumed to occur over the longer term. As Table ES-3 and Figure ES-2 show, replacing 25% of gasoline use with ethanol produced from cellulosic biomass could reduce both tailpipe and fuel-cycle GHG emissions by well over 20%. However, achieving these reductions would require developing and commercializing technology for producing ethanol from wood or herbaceous biomass rather than from corn, the current feedstock. Replacing one-quarter of projected gasoline use with compressed natural gas would reduce tailpipe and fuel cycle GHG emissions by 4% and 5%. Using liquid petroleum gas (LPG, or propane) as a gasoline substitute would reduce GHG emissions from vehicle use by only about 2%, as Table ES-3 and Figure ES-2 show; however, full fuel cycle emissions would be reduced by 5%, due to the low emissions that occur during LPG production and distribution compared to gasoline.

Petroleum-based diesel would reduce emissions by somewhat more – 6% in vehicle operations and 8% over the complete fuel cycle – if it could substitute for one-quarter of projected gasoline use by 2025. Blending petroleum diesel with a 20% soy-based component would increase the reduction in tailpipe GHG emissions to 9%, but larger energy demands and GHG emissions for producing the soy-based component would limit the reduction in fuel cycle GHG emissions to the same figure (8%) as for petroleum diesel. Battery-powered electric vehicles would again eliminate tailpipe emissions, so that replacing 25% of gasoline vehicle travel by electric vehicles would reduce tailpipe emissions by exactly that same percentage. While GHG emissions from generating electricity would erase some of this advantage, it would nevertheless remain significant -- 11%, as Table ES-3 reports and Figure ES-2 illustrates -- when measured over the full fuel cycle. Because it does not rely on carbon as a source of stored energy, using hydrogen as a fuel in internal combustion engines would also virtually eliminate tailpipe GHG emissions, although energy use and emissions in producing hydrogen from natural gas would reduce this advantage over gasoline to 14% when measured over the complete fuel production and use cycle.

While this study assumes that internal-combustion engines would continue to be the dominant technology used for light-duty vehicles, potential innovations in engine and drive system technologies could reduce GHG emissions significantly even if conventional fuels remained dominant, and by still larger amounts if their widespread use could be combined with the substitution of some alternative fuels for gasoline. As Table ES-4 indicates, direct-injection engines operating on gasoline or diesel could reduce GHG emissions per mile by about 20-25% in vehicle operation and by 20-30% over the complete fuel cycle, while operating these engines on ethanol derived from cellulosic biomass could reduce per-mile GHG emissions in vehicle operation and over the fuel cycle by as much as 80%. Of course, the reductions in *total* GHG emissions would also depend how widely these advanced engine technologies and alternatives to gasoline were used, but as the table shows, the same 25% replacement of conventional technology assumed to be possible for alternative fuels

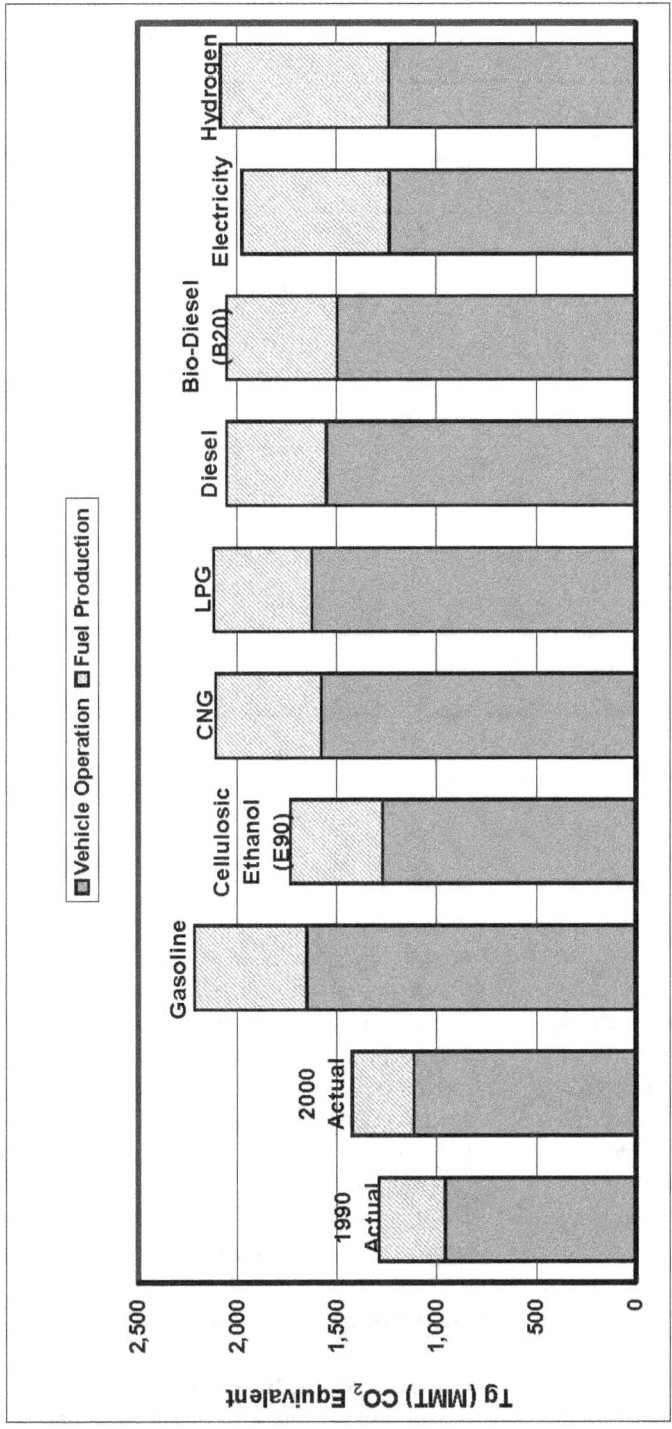

Figure ES-2. Estimated 2025 GHG emissions from light-duty vehicles with exclusive gasoline use and 25% alternative fuel replacement.

Table ES-3. Year 2025 greenhouse gas emissions with all-gasoline baseline and 25% gasoline replacement by alternative fuels.

Fuel Mix	Total GHG Emissions (Tg CO$_2$ equivalent) (1)		Reduction from Baseline (Tg)		% change from Gasoline Baseline	
	Vehicle Operation	Full Fuel Cycle	Vehicle Operation	Full Fuel Cycle	Vehicle Operation	Full Fuel Cycle
100% Gasoline	1,648	2,217	--	--	--	--
25% Ethanol (as E90), 75% Gasoline	1,271	1,731	377	487	-23%	-22%
25% CNG, 75% Gasoline	1,577	2,107	71	111	-4%	-5%
25% LPG, 75% Gasoline	1,623	2,117	25	100	-2%	-5%
25% Diesel, 75% Gasoline	1,551	2,051	97	167	-6%	-8%
25% Bio-Diesel (B20), 75% Gasoline	1,495	2,051	153	167	-9%	-8%
25% Electricity, 75% Gasoline	1,236	1,975	412	243	-25%	-11%
25% Hydrogen, 75% Gasoline	1,238	2,081	411	136	-25%	-6%

(1) Teragrams of CO$_2$ equivalent, computed using 100-year GWPs of 21 for CH$_4$ and 310 for N$_2$O. One teragram equals 10^{12} grams, or one million metric tons.

would reduce total emissions by 5-8% with gasoline or diesel as a fuel, and by as much as 20% if they were fueled by ethanol.

Still larger reductions in energy use – as much as 45-55% on a per-mile basis, as Table ES-4 also reports -- could result if light-duty vehicles were powered by hybrid internal combustion/electric drive systems using gasoline or diesel, or by fuel cells in vehicles employing on-board reforming of gasoline or methanol to produce hydrogen. Even larger reductions, between 58% and 87%, could be attained by ethanol-powered hybrid vehicles or by fuel cell vehicles using hydrogen produced at fueling stations from natural gas and stored aboard vehicles. Again, the resulting reductions in *total* GHG emissions would also depend on the fraction of conventional gasoline vehicles that could be replaced by these advanced-technology vehicle and fuel combinations, and the prospects for commercialization and widespread purchases of these extremely advanced technologies within the time horizon of this study remain at least as uncertain as those for most alternative fuels. However, very significant declines in total GHG emissions by light-duty vehicles could result if these vehicles accounted for a significant share of light-duty vehicle ownership and travel, particularly if many of them operated on substitutes for gasoline.

Criteria Pollutant Emissions

Federal and California regulations impose strict limits on light-duty vehicles' per-mile emissions of "criteria" air pollutants and their chemical precursors, including carbon monoxide (CO),

Table ES-4. Reductions in GHG emissions from alternative fuels and advanced engine technologies (% changes relative to gasoline in conventional ICE).

Engine Technology (1)	Fuel	GHG Emissions per Vehicle-Mile		2025 Total GHG Emissions with 25% Use	
		Vehicle Operation	Full Fuel Cycle	Vehicle Operation	Full Fuel Cycle
Conventional ICE	Gasoline	--	--	--	--
Direct Injection ICE	Gasoline	-19%	-20%	-5%	-5%
	Diesel	-24%	-30%	-6%	-8%
	Ethanol (E90)	-84%	-81%	-21%	-20%
Electric Hybrid	Gasoline	-46%	-46%	-12%	-12%
	Diesel	-50%	-54%	-12%	-13%
	Ethanol (E90)	-88%	-87%	-22%	-22%
Fuel Cell	Gasoline	-51%	-51%	-13%	-13%
	Methanol	-56%	-55%	-14%	-14%
	Hydrogen	-100%	-58%	-25%	-14%

(1) ICE indicates internal combustion engine.

volatile organic compounds (VOCs, which react in the atmosphere to produce ozone), oxides of nitrogen (NO_x, which forms ozone as well as other pollutants), and small particulate matter (PM_{10}, comprised of particles less than 10 microns in diameter). Progressive tightening of these emission standards over the past three decades has been a major source of improvements in air quality, so it is important that potential reductions in GHG emissions from substituting alternative fuels for gasoline be attainable without increasing emissions of these regulated pollutants or causing vehicles operating on them to violate current and pending emission standards.

The major concern with potential alternatives to gasoline appears to be higher NO_x and PM_{10} emission rates estimated by GREET for light-duty vehicles operating on petroleum-based diesel, since this is one of the non-gasoline fuels that appears capable of reducing GHG emissions significantly. The reduction in diesel fuel's sulfur content required by recently-adopted EPA regulations (scheduled to take effect beginning in 2006) is partly intended to facilitate use of exhaust aftertreatment devices that reduce NO_x and particulate emissions by diesel vehicles, a strategy that has proven extremely effective for gasoline engines. Nevertheless, some controversy remains about whether future light-duty diesel vehicles will be able to comply with future emissions standards for these pollutants even with the sharply reduced sulfur content permitted for diesel fuel.[4]

Table ES-5. Cost-effectiveness of near-term alternative fuels in reducing GHG emissions from light-duty vehicles.

Fuel	Incremental Capital Cost per Vehicle (2000 $)		Annual Incremental Fuel Cost per Vehicle (2000 $)	Annual GHG Reduction (Tg CO_2 equivalent) (2)	Cost/Ton of Emissions Avoided (2000 $) (3)
	Vehicle Production	Fuel Infra-structure (1)			
Ethanol (E85)	$284	$128	$604	51	$630
CNG (Bi-Fuel Vehicles)	$2,104	$398	-$66	8	$1,380
CNG (Dedicated Vehicles)	$1,914	$398	-$82	13	$310
LPG	$741	$222	$228	19	$430
Diesel	$2,500	$34	-$195	43	$50
Bio-Diesel (B20)	$2,500	$34	-$119	39	$100
Electricity	$4,100	$120	-$580	47	-$60

(1) Includes estimated investments for necessary storage, distribution, and retailing infrastructure.
(2) Reduction over Full Fuel Cycle; see Table ES-2. One Teragram (10^{12} grams) equals one million metric tons.
(3) Equals annualized total cost divided by annual reduction in GHG emissions in metric tons.

Cost-Effectiveness of Gasoline Substitutes

Total costs for replacing gasoline use with a substitute fuel include those for producing vehicles with the engine technology and on-board fuel storage capability allowing them to operate on a different fuel, costs for necessary investments in production, distribution, and retailing facilities for the substitute fuel, and costs for producing the substitute fuel itself, including those for feedstock production and fuel refining. Table ES-5 reports estimates of these three cost components for the fuels identified in this study as potential near-term substitutes for gasoline. These costs are measured *relative to those for gasoline*, and are expressed on a per-vehicle basis to facilitate their understanding and comparison.[5]

The number of light-duty vehicles projected to be in use during 2010 was used in conjunction with the incremental per-vehicle production costs and fuel infrastructure investments reported in Table ES-5 to estimate the total investment required to substitute each alternative fuel for 10% of projected gasoline use during 2010. These figures were then combined with assumptions about interest rates and typical lifetimes for vehicles and fuel production and distribution facilities to estimate the annualized equivalent values of these investments.[6] Next, the per-vehicle estimates of annual fuel cost differences between gasoline and each alternative fuel were multiplied by the number of gasoline vehicles that would be replaced to determine the total additional cost (or in some cases, savings) from substituting that fuel for 10% of gasoline use. Finally, the annualized incremental costs for vehicle production, fuel infrastructure, and fuel production were combined and divided by the previously-reported reductions in annual GHG emissions from each replacement fuel to determine its total cost per ton of emissions reduced.

These calculations suggest that replacing gasoline with most alternative fuels would be a costly strategy for reducing GHG emissions. As Table ES-5 reports, using petroleum-based diesel as a gasoline substitute appears likely to reduce GHG emissions at a cost of about $50 per ton, mainly because cost savings in refining diesel partly offset significantly higher costs for producing vehicles capable of using it as a fuel. Even this figure, however, compares unfavorably to many estimates of the cost-effectiveness of strategies for reducing GHG emissions from non-transportation sources, as well as to most recent estimates of the damage costs imposed by GHG emissions. Table ES-5 shows that other potential gasoline substitutes are likely to be still more costly ways to reduce GHG emissions: estimated costs per ton of GHG emissions avoided range from $100 for bio-diesel to more than $600 for corn-based ethanol.

Managing a Transition to Non-Gasoline Fuels

Substituting some alternative fuels for a significant fraction of gasoline use would require extremely large increases in their current and projected future production volumes. Correspondingly large increases in feedstock production volumes would also be required for gasoline substitutes that would be derived from feedstocks other than those now in widespread use, which are limited to petroleum and natural gas. These significantly higher levels of feedstock and fuel production, fuel distribution, and use of gasoline alternatives are likely to present some major concerns and required an extended period to achieve. For example, producing the quantity of cellulosic biomass necessary to support high-volume ethanol production would require devoting a significant amount of land to its cultivation, either by extensive displacement of current agricultural crops or a major expansion of total cultivated land.

A ubiquitous retailing or other distribution infrastructure comparable to that now offered by retail gasoline stations would also be required to allow any alternative fuel to replace a significant fraction of gasoline use. Refueling with any alternative fuel must approach the convenience now offered by gasoline if it is to become a widely-utilized replacement, including broad geographic availability, speed and simplicity of refueling, convenient hours of operation, availability of ancillary services such as vehicle repair, and options for self service. Refueling facilities would need to be available in a range of locations, including densely-developed urban areas, remote recreational locations, and at occasional points along highways in sparsely populated regions. While some alternative fuels such as ethanol and diesel could make extensive use of gasoline distribution and fueling infrastructure already in place with minor modifications, other fuels would require major investments to develop entirely new facilities for their distribution, storage, and retailing.

Both conventional and substitute fuels also present safety risks, potential hazards to human health, and possible environmental damages arising from their production, storage, and distribution, from normal vehicle operation and refueling, and from accidental fuel releases. Gasoline and alternative fuels are each characterized by different potential risks in their production, storage, distribution, and use, and the physical and chemical properties that create these risks vary considerably among fuels. The risk profiles presented by different fuels indicate that conventional fuels such as gasoline are not free from hazards, but they also show that most alternative fuel are not inherently more "dangerous" than the fuels they would replace.

The existence of special safety concerns or particular hazards with particular alternative fuels means that their safe and successful use will require intelligent planning of systems for their production and delivery. This includes careful design and engineering of facilities for their production, storage, and distribution, as well as of a network of outlets for refueling the large numbers of vehicles that would operate on them. Planning for their safe use must emphasize using these efforts to eliminate obvious hazards in their routine production and delivery, as well as to minimize the probability and consequences of lower-probability but more disastrous incidents, since even their rare occurrence can affect the use and acceptance of non-traditional fuels. Finally, thorough training in the handling and use of alternative fuels by both workers involved in their production and distribution, and by vehicle owners using these fuels, will be required to manage the potential hazards posed by their large-scale use.

Prospects for Alternative Fuels

Only a few alternative fuels appear to offer the potential to reduce GHG emissions from light-duty vehicle significantly from the levels generated by continued reliance on gasoline, regardless of the predominant vehicle and engine technology in use. Within the near-term time horizon of this study (approximately a decade), wider use of diesel, bio-diesel, corn-based ethanol, and electricity as light-duty vehicle fuels each appear to offer the potential for modest reductions in GHG emissions at realistic levels of gasoline replacement. Among these gasoline substitutes, however, only petroleum diesel appears likely to be cost-effective by comparison other strategies for reducing GHG emissions, and important concerns remain about whether light-duty diesel vehicles can comply with pending federal emission standards for certain criteria pollutants.

Over the longer term (20-25 years), ethanol derived from cellulosic biomass appears to offer the most realistic potential to achieve significant reductions in GHG emissions, because carbon emissions from its use as a fuel would be offset by carbon sequestration during growth of the biomass feedstock. However, substituting ethanol for a significant fraction of gasoline use would require successful large-scale commercialization of currently experimental technology to produce it from herbaceous or woody biomass, as well as major investments in production, storage, and distribution facilities. Because it does not rely on carbon as a source of stored energy, hydrogen offers the potential to eliminate GHG emissions in vehicle operation (some emissions would still be generated in producing and transporting it), but its commercial-scale production and distribution within even the year-2025 long term time horizon used in this study remains highly uncertain.

Advanced engine and drive system technologies – including direct injection internal combustion engines, hybrid internal combustion/electric drive systems, and fuel cells -- could also significantly reduce GHG emissions from their use over the longer term, by improving their energy efficiency and thus reducing total fuel energy consumption for light-duty vehicle travel. In combination with replacing a significant share of gasoline with certain alternative fuels, these advanced technologies offer the potential for very significant reductions in future GHG emissions compared to the level that would result from continued reliance on conventional internal combustion engines powered by gasoline.

However, costs for replacing a significant fraction of gasoline use with any alternative fuel – including additional costs for manufacturing vehicles, building or expanding facilities to produce, store, and distribute it, and producing the fuel itself -- are likely to be very substantial. Because substituting most other fuels for gasoline fuels appears likely to be a costly strategy for reducing GHG emissions from motor vehicle use, the desirability of government policies promoting their use depends critically on the range of *other* measures that are available to reduce GHG emissions -- including strategies to reduce emissions from the non-transportation sectors of the U.S. economy – and on how the costs of these competing measures compare to those for producing and using alternative fuels.

Fuel Options for Reducing Greenhouse Gas Emissions from Motor Vehicles

Energy Use and Greenhouse Gas Emissions

The transportation sector of the U.S. economy accounts for a major share of U.S. energy use. In 2001, transportation activity accounted for 28% of the nation's total energy consumption, and was responsible for nearly two-thirds of all *petroleum* energy consumed. In turn, the largest share – slightly more than three-quarters – of transportation energy use and petroleum consumption is accounted for by highway vehicles, including cars, light trucks used for personal and commercial travel, heavy-duty trucks, and buses.[7] Because a significant fraction of U.S. petroleum is imported, the transportation sector's heavy dependence on petroleum-based fuels may expose the nation's economy to the risk of potential price shocks or supply disruptions.

The nation's transportation sector also accounts for a significant share of greenhouse gas (GHG) emissions, which are widely believed to have the potential to alter the earth's climate. Highway vehicle use accounted for 24% of total 1999 U.S. emissions of carbon dioxide (CO_2), the most common greenhouse gas, and for a similar percentage of all GHGs when the different gases are weighted to reflect their potential contributions to climate change. Between 1990 and 2000, greenhouse gas emissions by motor vehicles grew at an annual rate of 2.1%, considerably faster than the 1.3% annual growth in emissions of GHGs from all sources.[8]

Finally, although air pollutant emissions from the transportation sector have declined significantly over the past two decades, transportation activity continues to account for significant fractions of certain pollutants. During 2000, motor vehicles were responsible for 55% of nationwide emissions of carbon monoxide (CO), as well as for 27% of volatile organic compounds (VOCs) and 36% of nitrogen oxides (NOx), both of which contribute to the unhealthful concentrations of ozone that occur in some U.S. urban areas.[9] Transportation activity also contributes a significant fraction of nationwide emissions of fine particulate matter, which is increasingly thought to represent a serious health hazard.

This report explores the potential of substitutes for gasoline to reduce energy use, GHG emissions, and air pollutant emissions from the use of light-duty motor vehicles in the U.S. These include automobiles as well as light-duty trucks – pickup trucks, vans, and sport/utility vehicles – that are increasingly used as substitutes for automobiles. It examines potential changes in energy use and emissions from using substitutes for gasoline to operate motor vehicles, as well as from producing and distributing substitute fuels and the feedstocks from which they are derived. The report also provides estimates of the costs of producing substitute fuels and modifying vehicles and fuel distribution infrastructure to allow them to replace some gasoline use, and uses these estimates to calculate the cost-effectiveness of substitute fuels as GHG control strategies.

What Are Alternative Fuels?

While substitutes for gasoline are commonly referred to as "alternative" fuels, there is no universally-recognized list of such fuels. The federal Alternative Motor Fuels Act of 1988 originally defined alternative fuels as limited to the alcohol fuels methanol and ethanol, as well as natural gas in either compressed or liquefied form. The 1992 Energy Policy Act (EPACT) expanded the definition of alternative fuels for light-duty vehicles to include mixtures containing 85% or more of methanol or ethanol, compressed or liquefied natural gas, liquefied petroleum gas (LPG), hydrogen, liquid fuels derived from coal, electricity, fuels produced from biomass, and others that are "substantially not petroleum."[10]

This study uses a similar definition, with the important exception that diesel fuels derived from petroleum and partly or completely from biomass (termed "bio-diesel") are included as potential substitutes for gasoline as light-duty vehicle fuels. Diesel fuels are included as alternatives to gasoline because this study emphasizes reductions in GHG emissions, and both petroleum-based diesel and bio-diesel have properties that offer the potential to reduce these emissions from the levels generated by the production and use of gasoline. These properties include the higher energy efficiency of compression-ignition diesel engines compared to spark-ignition engines, lower emissions of non-CO_2 GHGs from diesel fuel combustion, and lower GHG emissions in fuel refining and distribution compared to gasoline. Although the inclusion of diesel fuels represents an important departure from the statutory definition of the term, this study continues to refer to potential substitutes for gasoline that could be used by light-duty vehicles -- including petroleum-based diesel and bio-diesel -- as alternative fuels or potential substitutes for gasoline.

Why the Interest in Alternative Fuels?

Interest in substitutes for petroleum-based transportation fuels was originally motivated by concerns about the U.S. economy's vulnerability to disruptions in the supply of imported oil and abrupt increases in international petroleum prices, both of which were first experienced during the 1970s. Because transportation activity accounts for such a large share of petroleum use, reducing transportation's dependence on fuels derived from petroleum has the potential to reduce the nation's vulnerability to such events.

With heightened concern about urban air quality during the 1980s, the potential of alternative fuels to reduce motor vehicles' emissions of air pollutants attracted considerable interest. As the number of urban areas violating federal air quality standards declined and the limited potential of most alternative fuels to reduce air pollutant emissions became clear, however, their use began to appear less attractive. More recently, growing international pressure on the U.S. to reduce its contribution to potential climate change, together with renewed concerns about the reliability and pricing of petroleum imports, has rekindled interest in the potential of alternative fuels to reduce energy use and accompanying emissions of GHGs.

How Alternative Fuels Can Reduce GHG Emissions

Greenhouse gas emissions generated during the production and use of transportation fuels – including gasoline as well as most alternative fuels – consist mainly of carbon dioxide (CO_2), but also include trace amounts of methane (CH_4) and nitric oxide (N_2O). The "global warming potentials" of these latter two gases, which measure the potential contribution to long-term increases in atmospheric temperature from a given mass of each gas relative to that of the same amount of CO_2, are quite large (21 for CH_4 and 310 for N_2O). However, these gases are emitted in such small amounts that 97% or more of the potential contribution to global climate change from the production and use of transportation fuels results from carbon dioxide emissions.

There are several ways that substitutes for conventional fuels can reduce emissions of total GHGs from the use of motor vehicles or from fuel production:

- Fuels that can be used in conjunction with engine technologies that are more efficient in converting their stored energy into vehicle propulsion energy (for example, diesel or electricity) can reduce total transportation energy use and resulting GHG emissions.

- Fuels that contain less carbon per unit of stored energy (such as natural gas or – an extreme case -- hydrogen) can reduce emissions of CO_2, a combustion by-product of their stored carbon that accounts for most GHG emissions from transportation.

- Fuels that entail lower GHG emissions in their production, storage, and distribution can reduce total emissions over the "full fuel cycle," which extends from feedstock extraction through fuel production, storage, distribution, and use. (For example, carbon dioxide absorbed during the growth of biomass used to derive ethanol and bio-diesel offsets the carbon emissions that occur during subsequent combustion of those fuels.)

- Fuels that generate lower emissions of methane and nitric oxide in their production or use (such as electricity) can reduce total GHG emissions from transportation, although this potential is limited because these GHGs are emitted in such small amounts by production and use of most transportation fuels.

It is important to note that under international conventions for GHG emissions accounting, only emissions from fuel use during vehicle operation are assigned to the transportation sector. GHG emissions that occur during the extraction of fuel feedstocks, fuel refining, storage, and distribution are classified as industrial rather than transportation emissions. Nevertheless, fuels that generate lower GHG emissions in these "upstream" activities can reduce combined emissions of GHGs from all sources, and their use represents a potentially important strategy for reducing the nation's contribution to the threat of global climate change. As a consequence, this study focuses on GHG emissions occurring over the full fuel cycle, including feedstock production, refining and distribution, and fuel consumption by motor vehicles.

Scope of this Report

This report provides detailed empirical estimates of potential reductions in energy use and GHG emissions from a range of alternative fuels for light-duty vehicles, including automobiles and light-duty trucks, which are increasingly used for passenger travel. Separate estimates of GHG reductions from fuel use during vehicle operation alone, and over the complete fuel production and use cycle -- including feedstock extraction, fuel refining, distribution, and use -- are reported under the assumption that each alternative fuel could replace a significant fraction of projected future gasoline use. Estimates of GHG and criteria pollutant emissions were developed using the Greenhouse Gas and Regulated Emissions in Transportation (GREET) model developed by researchers at the U.S. Department of Energy's Argonne National Laboratories. The estimates of future GHG emissions and the potential of alternative fuels to reduce them reflect assumptions about growth in travel and changes in the vehicle fleet developed by staff of the U.S. Department of Transportation's Center for Climate Change and Environmental Forecasting.

Separate estimates of potential reductions in GHG emissions are developed for the near term, defined as approximately the year 2010, and the more distant future, defined as approximately the year 2025. These estimates are based on clearly identified assumptions about the potential extent to which various fuels could substitute for gasoline within each of those time horizons. They also reflect specific assumptions about light-duty vehicle technologies that are likely to be commercially available in the foreseeable future, the compatibility of these engine technologies with specific fuels, and the extent to which new engine technologies are likely to be represented in the nation's vehicle fleet by each of those dates.

The report also assesses "criteria" air pollutant emissions associated with each alternative fuel and engine technology combination for their compliance with adopted and pending emissions standards, a critical constraint on widespread production and use of new fuels. In addition, it estimates costs for producing and distributing the volume of each alternative fuel required by its assumed level of use during the near term, as well as for adapting vehicles to accommodate its on-board storage and use. These estimated costs are used in conjunction with the calculated GHG emission reductions from using each alternative fuel to evaluate its cost-effectiveness as a strategy to reduce GHG emissions within the study's near-term horizon.

Finally, the report identifies and briefly considers some potential concerns raised by the large-scale production, distribution, and use of each alternative fuel. These include possible safety and health concerns from production, storage, and consumer handling of various fuels, barriers to developing the infrastructure required to support commercial-scale production and distribution of each fuel, and potential limitations on production or extraction of the feedstock volumes necessary to produce various alternative fuels.

Critical Assumptions

Because the time horizon adopted for this study extends well into an uncertain future, assumptions about growth in motor vehicle use, the number and types of vehicles in service, and the engine technologies and fuels they use must be made to establish a "baseline" against which to assess the potential of substitute fuels to reduce GHG emissions. There is also considerable uncertainty about future developments in fuel supplies and vehicle technologies, so specific assumptions about the sources and properties of different fuels and the availability of compatible engine technologies are also required to enable precise estimates of GHG emissions reductions. This section discusses the key assumptions about motor vehicle use, fuel availability, and future developments in engine technology that were made as part of this study. Table 1 summarizes these assumptions.

It is important to note that these assumptions are *not* intended as forecasts of vehicle use, developments in vehicle technology, or displacement of gasoline by alternative fuels. Instead, they are assumptions made partly for the purpose of reducing the effects of uncertainty about future trends in these variables on estimates of future GHG emissions levels under baseline or "business as usual" conditions. These assumptions are also intended to highlight the distinction between potential reductions in GHG emissions that would stem from realistic future levels of alternative fuel use, and those that might result from innovations in vehicle technology. Since there are few cases where use of specific alternative fuels requires -- or even enables – the development and production of advanced engine technologies, this distinction is important to maintain.

Growth in Motor Vehicle Use

The number of miles driven in light-duty vehicles (LDVs, including automobiles and light-duty trucks) is assumed to grow by 2.5% annually through the year 2010, the near-term horizon adopted for this study. After that date, growth in light-duty vehicle use is projected to slow to 2.0% annually through the year 2025, the study's long-term time horizon. As a result, total light-duty vehicle travel is projected to rise from the 2.34 trillion miles recorded during 2000 to nearly 3 trillion miles by 2010, and to slightly more than 4 trillion miles by 2025.[11]

Light-duty trucks used primarily for personal transportation are expected to represent a growing fraction of all light-duty vehicle

Table 1. Summary of key assumptions used in study.

Assumption	Date	Light-Duty Vehicles
VMT Growth	2000-2010	2.5% per year
	2010-2025	2.1% per year
VMT Mix	2000	Passenger Cars: 63% Small Light Trucks: 27% Large Light Trucks: 10%
	2010	Passenger Cars: 55% Small Light Trucks: 33% Large Light Trucks: 12%
	2025	Passenger Cars: 47% Small Light Trucks: 39% Large Light Trucks: 14%
Baseline Fuel	2010	Gasoline (70% Conventional, 30% Federal Phase 2 RFG) (1)
	2025	Gasoline (70% Conventional, 30% Federal Phase 2 RFG) (1)
Baseline Engine Technology	2010	Spark-Ignition Internal Combustion
	2025	Spark-Ignition Internal Combustion
Baseline Energy Efficiency Improvement (2)	2000-2010	Vehicle: 1.2% Full Fuel Cycle: 1.2%
	2010-2025	Vehicle: 4.0% Full Fuel Cycle: 3.8%
Alternative Fuel Use (% of VMT) (2)	2010	10%
	2025	25%

(1) Sulfur content reduced in compliance with provisions of EPA "Tier 2" rule on gasoline sulfur content.

(2) Average for all light-duty vehicle travel, consisting of assumed mix of vehicle classes shown above.

travel, increasing from an estimated 37% in 2000 to 45% by 2010, and further to 53% by the year 2025. As a consequence, conventional automobiles -- which accounted for 63% of all light-duty vehicle use during 2000 -- will represent less than half of LDV travel by 2025.[12] Smaller light truck models (defined as those under 6,000 pounds Gross Vehicle Weight) are expected to continue to account for two-thirds or more of light truck use throughout this period, as Table 1 indicates.

Fuel Use and Engine Technology

Without specific efforts to promote development and use of alternative fuels, the U.S. light-duty vehicle fleet is likely to continue to operate predominantly on gasoline. This study assumes that the current mix of approximately 70% conventional and 30% "reformulated" gasoline will continue to be used in the absence of such efforts.[13] It also assumes that the 30 parts per million (ppm) limit on gasoline sulfur content included as part of the Environmental Protection Agency's (EPA) "Tier 2" light-duty vehicle emission standard will be fully implemented by 2007 as scheduled.[14] EPA's recently-adopted rule limiting the sulfur content of diesel fuel to 15 ppm is also assumed to take effect as scheduled in 2006.[15]

Near-Term Assumptions

Conventional internal-combustion engine technology is expected to continue to dominate the light-duty vehicle fleet through this study's future horizon. While "evolutionary" improvements in the efficiency of internal-combustion engines and vehicle drive trains are expected to continue over this period, thus offering the potential for improved fuel economy, consumer demands for increases in vehicle size and performance are expected to absorb much of this efficiency improvement. At the same time, continued substitution of less fuel-efficient light-duty trucks for automobiles is also likely to offset some of the potential improvement in vehicle efficiency. As a consequence, Table 1 indicates the overall fuel efficiency of the U.S. light-duty vehicle fleet is expected to improve by only about 1% through 2010, and 5% by 2025.

The assumptions about potential use of substitute fuels adopted for this study are intended to represent reasonable estimates of the displacement of gasoline that could result from concerted efforts by government to encourage their production and use. By the study's 2010 near-term horizon, financial incentives or other policies promoting substitutes for gasoline are assumed to have the potential to replace 10% of projected gasoline use by light-duty vehicles. For substitute fuels that require specific engine technologies or specially-designed fuel delivery systems, this assumption implies that a corresponding fraction of the light-duty vehicle fleet – including both automobiles and light trucks -- would consist of vehicles capable of operating on those fuels by the year 2010.

However, alternative fuels that are suitable for use in vehicles designed to operate on gasoline, or require only minimal modifications to conventional gasoline vehicles, might achieve this target without requiring specially-designed vehicles to reach such a large share of the light-duty fleet. One example is ethanol, for which the 10% displacement target could be achieved simply by blending it in that proportion with gasoline for universal use by conventional light-duty vehicles. Natural gas could also reach the 10% use target if vehicles capable of operating on gasoline as well as on natural gas (called bi-fuel vehicles) achieved the necessary representation in the fleet.[16] The estimates of changes in energy use and GHG emissions reported for each alternative fuel assume that all of the assumed reduction in gasoline use is replaced by that fuel, while all remaining light-duty vehicle travel continues to be powered by gasoline.

Assumptions About Longer-Term Developments

Over this study's longer-term horizon, various innovations in light-duty engine and vehicle technology could produce significant improvements in energy efficiency, thus offering the potential to reduce GHG emissions even without the replacement of gasoline by alternative fuels. Gasoline and diesel direct injection engines, currently in use on some vehicles sold outside the U.S., could offer a significant improvement over conventional internal combustion engines if challenges in controlling their emissions of certain pollutants can be overcome. Hybrid electric drive systems, which combine a small internal-combustion engine with an on-board storage battery and a battery-powered electric motor, are also likely to become commercially available in a range of light-duty vehicle models within this study's longer-term time horizon. Hybrid-drive vehicles are expected to offer a significant energy efficiency advantage over internal combustion engines, although the extent to which they are likely to enter the light-duty fleet is uncertain, primarily because costs for producing them and future gasoline prices – two major influences on vehicle buyers' decisions – remain speculative. Similarly, fuel cell technology may offer the potential for even greater increases in vehicle efficiency, but correspondingly larger technological and economic uncertainty also surrounds the commercial development of fuel cell-powered vehicles.[17]

Thus it is extremely important to distinguish between *differences* in projected future GHG emissions levels resulting from different assumptions about how widely advances in engine technology will be incorporated into the vehicle fleet, and *reductions* in emissions likely to result from operating the mix of conventional and advanced engine technologies assumed to comprise the future vehicle fleet on various substitute fuels instead of on gasoline. While this study devotes some attention to the potential reductions in gasoline use and emissions that could result from commercial development of advanced engine and vehicle technologies, it clearly distinguishes the contribution of alternative fuels *themselves* to reducing GHG emissions from the effect of advances in vehicle technology. The effect of substituting alternative fuels for gasoline, which is the major focus of this study, is measured by the difference in emissions for the *same* vehicle fleet when operating exclusively on gasoline, and on a mix of gasoline and an assumed fraction (10% in the near term and 25% in the longer term) of each alternative fuel.

Greenhouse Gas Reductions from Light-Duty Vehicles

Gasoline-powered motor vehicles -- primarily autos and light truck models used for personal travel -- currently account for nearly 60% of transportation-sector GHG emissions, and for about 20% of total U.S. emissions of GHGs from all sources.[18] Thus the potential for reducing U.S. GHG emissions by encouraging alternative fuel use by light-duty vehicles may be significant, especially for non-gasoline fuels that can improve the energy efficiency of light-duty vehicles or that derive less of their stored energy content from carbon.[19]

Producing and distributing motor vehicle fuels also entails substantial energy use and GHG emissions, and using fuels that generate lower levels of these "upstream" emissions is another potentially important way to lower GHG emissions from light-duty vehicle use. This section estimates potential reductions in total energy use and GHG emissions from alternatives to gasoline during the study's 2010 near-term horizon, while the following section addresses potential GHG reductions from more extensive use of alternative fuels and innovations in vehicle technology over the longer term.

How the Estimates Were Developed

This analysis focuses on fuels and engine technologies that appear to offer the most realistic potential for widespread production and use within the 10 and 25-year time horizons. Estimates of energy use and emissions per vehicle-mile of travel were developed using the Greenhouse Gases and Regulated Emissions from Transportation (GREET) model developed by researchers at Argonne National Laboratories, under the sponsorship of the U.S. Department of Energy. The model's estimates of energy use and emissions of GHGs and regulated air pollutants are based on extensive analysis of fuel efficiency and emissions testing data for light-duty vehicles drawn from government, industry, and other sources, supplemented with mass-balance calculations based on the energy and carbon content of various fuels.

GREET also calculates energy use from upstream fuel processing stages based on measured or estimated energy consumption rates and energy conversion efficiencies for each individual stage of fuel production and distribution, together with assumptions about the specific energy sources used in processing gasoline and other transportation fuels. It combines these estimates of energy use in fuel production and distribution with fuel-based emission factors to estimate upstream GHG and air pollutant emissions per unit of transportation fuel energy produced and consumed.[20]

The GREET model estimates separate energy consumption and emissions rates per vehicle-mile of travel for automobiles, small light trucks (those under 6,000 pounds Gross Vehicle Weight), and larger light trucks (6,000-8,550 pounds Gross Vehicle Weight) operating on gasoline and other fuels. Tables A-1 through A-6 of the Appendix report energy consumption, greenhouse gas, and criteria pollutant emission rates per vehicle-mile of travel for automobiles, small light trucks, and large light trucks during the years 2010 and 2025. These energy use and emission rates for the three individual vehicle classes (automobiles, small light trucks, and large light trucks) are weighted by the fractions of light-duty vehicle travel assumed to be accounted for by each class (reported previously in Table 1) to determine composite energy consumption and emission rates per vehicle-mile for the entire light-duty vehicle fleet during 2010 and 2025. The estimates of potential near- and longer-term savings in energy use and reductions in greenhouse gas emissions from using alternative fuels presented in this report are based on these composite energy use and emission rates, and thus reflect the expected composition of the U.S. vehicle fleet in 2010 and 2025.

Total energy use and emissions of each GHG or regulated air pollutant generated by light-duty vehicle use are estimated by multiplying these fleet-wide composite energy use and emissions rates per vehicle-mile of travel by the forecasts of total light-duty vehicle travel for the near- and long-term horizon years, also reported previously in Table 1. These calculations are first performed assuming that all light-duty vehicles operate on gasoline (using the mix of conventional and reformulated gasoline specified previously in Table 1) to establish baseline future levels of GHG emissions. The calculations are then repeated under the assumption that each alternative fuel replaces the assumed fraction of projected travel by gasoline-powered vehicles – 10% in the year 2010, and 25% by the year 2025 – to estimate the resulting levels of energy use and GHG emissions. Finally, the corresponding reductions in energy use and emissions are calculated as the *difference* between their levels with the assumed fraction of gasoline replacement by each alternative fuel and their levels under the all-gasoline baseline.

Greenhouse gas emissions estimated by GREET include carbon dioxide (CO_2), methane (CH_4), and nitric oxide (N_2O). Because different fuels generate varying mixes of these three gases, this study weights methane and nitric oxide by their carbon dioxide equivalents using estimated 100-year "Global Warming Potentials" (GWPs), and reports total GHG emissions in terms of their CO_2 mass equivalent. GHG emissions generated by the assumed levels of use of different alternative fuels are reported both as total CO_2-equivalent tons and as changes from the level of emissions projected to result from the all-gasoline baseline. Energy consumption and GHG emissions are reported separately for vehicle operation and for the "full fuel cycle," which adds energy use and emissions that occur during feedstock extraction and processing, fuel refining, storage, distribution, and vehicle re-fueling to those from fuel combustion during vehicle operation.

Near-Term Energy Use and GHG Emissions

Several alternative light-duty vehicle fuels are assumed to have reasonable prospects for commercial-scale production and use within the 2010 near-term time horizon adopted for this study:

- ethanol (blended with 15% gasoline, referred to as E85)

- compressed natural gas (CNG), which can be used in both bi-fuel (vehicles that can operate on either gasoline or CNG) and dedicated vehicles (which operate only on CNG)

- liquid petroleum gas (propane)

- conventional petroleum-based diesel and "bio-diesel," which blends petroleum-based diesel with a soy-based component

- electricity used in battery-powered electric vehicles.

Table 2 identifies the feedstock used to derive each fuel and the engine technology likely to be used by vehicles operating on it. With the exception of battery-powered electric vehicles, each of these fuels and vehicle technologies is in commercial-scale production today, although obviously at volumes far below gasoline and internal combustion engines designed to use it as a fuel.

Table 2. Energy efficiency of gasoline baseline and near-term alternative fuels for light-duty vehicles.

Fuel	Feedstock	Engine Technology (1)	Energy Efficiency (Btu/vehicle-mile)	
			Vehicle Operation	Full Fuel Cycle
Gasoline	Petroleum	Spark-Ignition ICE	6,065	7,659
Ethanol (E85)	Corn (85%)/ Petroleum (15%)	Spark-Ignition ICE: Flexible-Fuel Vehicle	5,822	9,071
CNG	Natural Gas	Spark-Ignition ICE: Bi-Fuel Vehicle	6,739	8,293
CNG	Natural Gas	Spark-Ignition ICE: Dedicated Vehicle	6,522	8,026
LPG	Petroleum/ Natural Gas	Spark-Ignition ICE: Dedicated Vehicle	6,065	6,936
Diesel	Petroleum	Compression-Ignition ICE	4,493	5,372
Bio-Diesel (B20)	Petroleum (80%)/ Soy (20%)	Compression-Ignition ICE	4,493	5,562
Electricity	Projected U.S. Generating Mix	Battery-Powered Electric Motor	0	6,596

(1) ICE indicates internal combustion engine.

Source: Estimated using *GREET 1.5a -- Transportation Fuel-Cycle Model*, Argonne National Laboratories, January 2000, using assumptions reported in Table 1. See Appendix Table A-1 for energy consumption rates of individual light-duty vehicle classes.

Energy Efficiency with Alternative Fuels

Table 2 also reports the energy efficiency of each fuel and engine technology combination (measured in Btu consumed per vehicle-mile of travel) for the mix of light-duty vehicle classes expected to be in use during 2010 (reported previously in Table 1), both during vehicle operation itself and over the complete fuel production and use cycle.[21] Energy consumption per mile of travel is higher over the full fuel cycle than in vehicle operation alone -- often significantly -- because of the additional energy demands for extracting feedstocks, refining or producing, and storing and distributing many fuels. The energy efficiency figure for light-duty vehicles powered by internal combustion engines operating on gasoline in the first row of Table 2 represents the baseline against which other fuels used in the same type of engine (and electricity used by battery-powered electric vehicles) are compared.

As the table indicates, replacing gasoline with ethanol (E85) improves energy efficiency in vehicle operation slightly, but reduces it significantly when evaluated on a full fuel cycle basis because of the energy intensity of producing the required feedstock (corn, in the near term) and processing it into motor fuel. Compressed natural gas reduces the energy efficiency of vehicle operation both in dedicated and bi-fuel vehicles, because of the slightly lower efficiency with which internal combustion engines are able to convert its stored energy content to propulsion energy, and the additional weight associated with bi-fuel vehicles' parallel fuel storage and delivery systems.

However, lower energy demands for producing CNG offsets part of this disadvantage, as shown by the smaller increases in energy consumption compared to the baseline gasoline vehicle over the full fuel cycle (8% for bi-fuel and 5% for dedicated vehicles) than in vehicle operation alone (11% and 8%). Liquefied petroleum gas (LPG), a by-product of petroleum refining and natural gas extraction, would not change energy use in vehicle operation compared to gasoline, but would improve energy efficiency by about 9% when fuel production *and* use are considered.

Other alternatives to gasoline appear to offer potentially significant energy efficiency improvements in both vehicle operations and over the complete fuel cycle. These include petroleum-based diesel, blends of petroleum-derived and soy-based diesel fuels (the data in Table 2 refer to an 80% petroleum/20% soy blend), and electricity.[22] For the mix of automobiles and light-duty trucks assumed to comprise the U.S. fleet in the near term, the superior energy conversion efficiency of compression-ignited (diesel) engines would reduce energy use in vehicle operation by 26% compared to gasoline, and the lower energy demand for refining petroleum diesel would increase the energy efficiency advantage of these fuels slightly over the fuel cycle.

Because they operate on stored energy, battery-powered electric vehicles eliminate energy consumption during vehicle operation, as the zero entry in Table 2 indicates. Using nationwide average values for energy efficiency in generating and transmitting electricity, however, the energy efficiency advantage of battery-powered electric vehicles over their gasoline-powered counterparts is reduced to about 14% when measured over the full fuel cycle.

GHG Emissions Reductions from Vehicle Operation

Partly as a consequence of their higher energy efficiency, some alternative fuels may offer the potential for significant reductions in GHG emissions from operating light-duty vehicles within this study's ten year near-term horizon. The higher energy efficiency or lower "carbon intensity" of some fuels compared to gasoline results in lower emissions per vehicle-mile of carbon dioxide (CO_2), the primary greenhouse gas generated by fuel combustion, while some alternative fuels also feature lower emissions of methane (CH_4) or nitric oxide (N_2O). As Table 3 shows, all of the potential substitutes for gasoline considered in this study reduce GHG emissions per mile of travel in light-duty vehicle operation, with ethanol, diesel fuels, and electricity each producing significant reductions.[23]

Ethanol's extremely low CO_2 emissions rate in vehicle operation occurs primarily because the GREET model credits carbon sequestration (the absorption of carbon dioxide) that occurs during corn growth against CO_2 emissions occurring during the *operation* of ethanol-fueled vehicles, rather than against other emissions that occur during ethanol *production*. As with energy use, GREET also calculates that electricity produces zero GHG emissions during vehicle use because no *additional* emissions are produced beyond those from generating the electrical energy that is stored in these vehicles' on-board batteries, and emissions produced in generating electricity are accounted for in the fuel production stage.

Alternative fuels that produce large reductions in GHG emissions *per mile* of travel can also lower *total* GHG emissions generated by the level of light-duty vehicle travel projected for 2010, although the modest targets for gasoline replacement assumed to be feasible by that date limit the potential reduction in total GHG emissions from using alternative fuels. Table 4 shows that total GHG emissions from vehicle operation with exclusive use of gasoline as a fuel for light-duty vehicles would total 1,373 teragrams (or million metric tons, measured in CO_2-equivalent terms) during that

Table 3. Greenhouse gas emissions rates for light-duty vehicles using gasoline and near-term alternative fuels.

Fuel	GHG Emissions Rates (grams/vehicle-mile)							
	Vehicle Operation				Full Fuel Cycle			
	CO_2 (1)	CH_4 (2)	N_2O (3)	Total (4)	CO_2 (1)	CH_4 (2)	N_2O (3)	Total (4)
Gasoline	449	0.085	0.031	460	558	0.814	0.039	588
Ethanol (E85)	99	0.130	0.031	111	365	0.669	0.206	443
CNG	402	0.867	0.019	426	504	2.332	0.020	559
CNG	389	0.867	0.025	414	488	2.284	0.026	544
LPG	434	0.113	0.031	446	498	0.762	0.032	524
Diesel	361	0.013	0.021	368	428	0.479	0.021	444
Bio-Diesel (B20)	367	0.013	0.021	374	441	0.385	0.024	457
Electricity	0	0.000	0.000	0	416	0.599	0.003	430

(1) Carbon dioxide; used as numeraire to express "Global Warming Potential" (GWP) of other GHGs.

(2) Methane; GWP relative to carbon dioxide = 21 (100-year value).

(3) Nitrous oxide; GWP relative to carbon dioxide = 310 (100-year value).

(4) CO_2 equivalent, with methane and nitrous oxide weighted by GWPs.

Source: Estimated using *GREET 1.5a -- Transportation Fuel-Cycle Model*, Argonne National Laboratories, January 2000, using assumptions reported in Table 1. See Appendix Table A-2 for greenhouse gas emission rates of individual light-duty vehicle classes.

year. Table 4 also shows total GHG emissions with each alternative fuel accounting for 10% of projected light-duty vehicle travel, while the remaining 90% continues to take place in gasoline-powered vehicles. Finally, the table reports the resulting reductions in total GHG emissions by light-duty vehicles compared to the all-gasoline baseline.

As Table 4 reports, replacing 10% of projected gasoline use with corn-based ethanol is estimated to reduce GHG emissions from operating light-duty vehicles nearly proportionally (8.9%), as a result of carbon sequestration during the growth of its corn feedstock. Much smaller reductions in tailpipe GHG emissions by light-duty vehicles (1-2%) could also result if CNG, petroleum diesel, or bio-diesel displaced 10% of gasoline use, while use of LPG would produce an even more modest reduction. Since GREET attributes energy use and GHG emissions associated with generating and transmitting electricity to fuel production rather than to actual operation of electric vehicles, it estimates that replacing 10% of gasoline vehicle travel by electric vehicles would produce a corresponding 10% reduction in total GHG emissions.

"Full Fuel Cycle" GHG Emissions

A more comprehensive measure of the impact on nationwide GHG emissions from using substitutes for gasoline is the resulting change in emissions measured over the full fuel cycle, including emissions that occur during fuel

Table 4. Year 2010 greenhouse gas emissions with all-gasoline baseline and 10% gasoline replacement.

Fuel Mix	Total GHG Emissions (Tg CO_2 equivalent) (1)		Change from Gasoline Baseline (Tg)		% change from Gasoline Baseline	
	Vehicle Operation	Full Fuel Cycle	Vehicle Operation	Full Fuel Cycle	Vehicle Operation	Full Fuel Cycle
100% Gasoline	1,373	1,752	--	--	--	--
10% Ethanol (as E85), 90% Gasoline	1,250	1,701	-122	-51	-8.9%	-2.9%
10% CNG (Bi-Fuel Vehicles), 90% Gasoline	1,362	1,743	-10	-8	-0.8%	-0.5%
10% CNG (Dedicated Vehicles), 90% Gasoline	1,359	1,739	-14	-13	-1.0%	-0.7%
10% LPG, 90% Gasoline	1,368	1,733	-4	-19	-0.3%	-1.1%
10% Diesel, 90% Gasoline	1,345	1,709	-28	-43	-2.0%	-2.4%
10% Bio-Diesel (B20), 90% Gasoline	1,347	1,713	-26	-39	-1.9%	-2.2%
10% Electricity, 90% Gasoline	1,235	1,705	-137	-47	-10.0%	-2.7%

(1) Teragrams of CO_2 equivalent, computed using 100-year GWPs of 21 for CH_4 and 310 for N_2O. One teragram equals 10^{12} grams, or one million metric tons.

Source: Calculated from Tables 1 and 3.

production and distribution as well as during vehicle operation. As with energy consumption, GHG emission rates over the full fuel cycle are higher for all fuels because of emissions occurring during fuel production and distribution. Nevertheless, Table 3 previously showed that full fuel cycle GHG emissions per vehicle-mile remain significantly lower for ethanol, diesel fuels, and electricity than for gasoline.

Table 4 uses these emission rates and the forecast of total vehicle use for 2010 to estimate total GHG emissions from light-duty vehicle travel for the all-gasoline base case, as well as for the case where each alternative fuel replaces 10% of gasoline use. As it shows, including GHG emissions from fuel production and distribution lowers the estimated emissions reduction from replacing 10% of gasoline use with corn-based ethanol to only about 3%, primarily as a result of the large energy demands and associated GHG emissions in growing and harvesting corn to provide the necessary feedstock. Including these "upstream" sources of emissions actually *increases* potential GHG reductions from petroleum-based diesel and bio-diesel slightly, because of lower energy use and emissions in refining crude petroleum into diesel fuel. The reduction in total GHG emissions with 10% of light-duty vehicle travel powered by electricity drops to less than 3% when typical energy losses in electricity generation and transmission are accounted for, as Table 4 also reports.

Criteria Pollutant Emissions from Alternative Fuels

Federal and California regulations impose strict limits on light-duty vehicles' per-mile emissions of various "criteria" air pollutants and their chemical precursors, including carbon monoxide (CO), volatile organic compounds (VOCs, which react in the atmosphere to produce ozone), oxides of nitrogen (NO_x, which form ozone as well as other pollutants), and small particulate matter (PM_{10}, comprised of particles less than 10 microns in diameter). Progressive tightening of these emission standards over the past three decades has been a major source of improvements in air quality, and it is important that the significant benefits to human health and property from tighter emission standards be sustained. Thus it is important to verify that potential reductions in GHG emissions from using alternative fuels can be attained without resulting in accompanying increases in emissions of regulated pollutants. Since vehicles are required to comply with emission standards regardless of the fuel on which they operate, it is also important to examine whether emissions of criteria pollutants represents a potential regulatory constraint on the use of alternative fuels that have the potential to reduce GHG emissions.

The GREET model produces estimates of average per-mile emissions of criteria pollutants for light-duty vehicles during actual on-road driving. Because new vehicles' compliance with federal and California emissions standards is tested under carefully controlled conditions that differ from those encountered in on-road driving, GREET's estimates of light-duty vehicles' emission rates when operating on alternative fuels cannot be used to test directly their compliance with emission standards. However, the model's estimated emission rates for light-duty vehicles using alternative fuels can be compared to those for the same vehicles operating on gasoline, in order to identify potential increases in air pollutant emissions that might result from using non-gasoline fuels.[24]

This comparison, which is reported in Table 5, indicates that most alternative fuels likely to be widely available within the study's near-term horizon would produce criteria pollutant emissions by light-duty vehicles that are no higher than those from using gasoline, and in many cases much lower.[25] According to GREET's emissions estimates, all of the substitute fuels considered would reduce VOC and CO emissions significantly compared to gasoline, and most would do so for NO_x and SO_x as well. However, petroleum diesel and biodiesel appear to increase emissions of NO_x, another ozone precursor, as well as of SO_x and PM_{10}.

Since the coarser component of PM_{10} emissions from light-duty vehicles consists primarily of tire fragments and road dust, which are unaffected by the specific fuel in use, higher PM_{10} emissions with diesel fuels probably reflects increased emissions of extremely fine airborne particles that -- while not now directly regulated -- are an increasing concern for human health. Increased emissions of NO_x and SO_x from diesel fuels compared to gasoline will also contribute indirectly to airborne levels of fine particulate matter, since some is a product of secondary particle formation from gaseous compounds that occurs in the atmosphere.

The potential increase in emissions of these pollutants from diesel fuels is a major concern, since these fuels appear to be potentially effective in reducing GHG emissions. The sharp reduction in diesel fuel's sulfur content required by recently-adopted EPA regulations is partly intended to facilitate use of aftertreatment devices that reduce NO_x and particulate emissions, so it is possible that light-duty diesel vehicles will be able to comply with the stricter limits on these pollutants that apply to future model years (the sulfur content restriction takes full effect in 2006). However, it is also possible that new evidence on adverse health effects from very small particulate matter will prompt the adoption of even more restrictive emission standards within this study's near-term time frame. Thus it is important that the capability of diesel vehicles to meet pending (and perhaps even stricter) emission standards be clearly demonstrated if diesel fuel is to offer a viable alternative to gasoline for reducing GHG emissions by light-duty vehicles.

Table 5. Criteria pollutant emission rates for gasoline and near-term alternative fuel/engine technology combinations.

Fuel	Emission Rates (grams/vehicle-mile)				
	VOC (1)	CO (2)	NO_x (3)	SO_x (4)	PM10 (5)
Gasoline	0.257	7.297	0.410	0.044	0.034
Ethanol (E85)	0.222	4.359	0.368	0.012	0.026
CNG (Bi-Fuel Vehicles)	0.144	6.011	0.416	0.002	0.022
CNG (Dedicated Vehicles)	0.064	5.811	0.389	0.002	0.022
LPG	0.124	5.822	0.368	0.000	0.022
Diesel	0.138	1.109	0.674	0.057	0.122
Bio-Diesel (B20)	(6)	(6)	(6)	(6)	(6)
Electricity	0.000	0.000	0.000	0.000	0.021

(1) Volatile organic compounds.
(2) Carbon monoxide.
(3) Nitrogen oxides.
(4) Sulfur oxides.
(5) Particulate matter less than 10 microns in diameter.
(6) Not available; assumed to be comparable to petroleum diesel.

Source: Estimated using *GREET 1.5a -- Transportation Fuel-Cycle Model*, Argonne National Laboratories, January 2000, using assumptions reported in Table 1. See Appendix Table A-3 for criteria pollutant emission rates of individual light-duty vehicle classes.

Longer-Term GHG Emissions Reductions from Alternative Fuels

Over the longer-term horizon adopted for this study (the year 2025), gradual improvements in refinery yields and related fuel production processes are likely to increase the availability and reduce the cost of producing currently available alternative fuels, while lowering upstream energy use and GHG emissions associated with their production. In addition, the commercialization of technology for producing fuel ethanol from cellulosic biomass rather than corn (the current feedstock) could expand ethanol supplies significantly, while also reducing its cost. Finally, commercial-scale production of some additional alternatives to gasoline, such as hydrogen derived from natural gas, might also become feasible.

At the same time, important innovations in vehicle and engine technology are likely to reach commercial-scale production, many of which have the potential to improve the energy efficiency of light-duty vehicles significantly. These technologies include spark-ignited and compression ignition internal combustion engines that employ direct fuel injection, hybrid internal combustion/electric drive systems using a small internal combustion engine to generate electricity that is stored in an on-board batteries and used to power an electric motor, and proton-exchange membrane fuel cells, which are now under experimental development. Direct-injection engines, hybrid electric drive systems, and perhaps even fuel cells could be produced in a range of vehicle models sufficiently in advance of the long-term horizon year that a significant share of light-duty vehicle travel might take place in vehicles using one of these technologies by 2025.[26]

Nevertheless, the prospects for successful development and commercial-scale production of these advanced technologies remain somewhat uncertain. Further, it is not clear whether vehicle buyers would purchase vehicles employing these technologies on a large scale without significant increases in gasoline prices or new regulatory requirements for manufacturers to sell more efficient vehicles. In order to respond to this uncertainty, this study first estimates potential longer-term reductions in GHG emissions from more extensive displacement of gasoline by substitute fuels under the assumption that internal combustion engines (with continued "evolutionary" improvements in their efficiency) remain the dominant technology in use. Specifically, each alternative fuel considered to be a viable gasoline substitute in the short term -- as well as hydrogen, which could become commercially available within the longer-term horizon -- is assumed to replace 25% of projected gasoline use by the year 2025, and the resulting reductions in energy use and GHG emissions are estimated.

The study next explores potential reductions in GHG emissions resulting from the use of advanced engine and vehicle technologies, including direct-injection internal combustion engines, hybrid electric drive systems, and fuel cell vehicles. In order to highlight the potential role of *engine technologies themselves* in reducing GHG emissions, per-mile emissions for vehicles powered by each of these technologies when operating on gasoline as a fuel are first compared to those for a conventional gasoline-powered internal combustion vehicle. Next, the incremental reduction in GHG emissions from the use of alternative fuels in conjunction with these advanced technologies is illustrated by comparing their per-mile emissions when operating on selected alternative fuels. Finally, estimates of reductions in total GHG emissions are presented assuming that travel by each of these advanced technology and fuel combinations accounts for the same 25% of total light-duty vehicle use during the year 2025 as assumed previously for alternative fuels.

Energy Efficiency with Alternative Fuels

After accounting for the increased share of the fleet expected to be made up of light trucks by 2025, the energy efficiency of light-duty vehicles powered by gasoline-fueled internal combustion engines is projected to increase approximately 5% from its current level. Table 6 compares the energy efficiency of this same vehicle mix (47% automobiles and 53% light trucks, as reported previously in Table 1) when operating on various alternative fuels to this slightly more efficient gasoline baseline.[27] As it shows, ethanol (blended with 10% gasoline, or E90) and LPG would each further improve energy efficiency in vehicle operation by about 8% compared to gasoline. However, extremely high energy use in producing ethanol from cellulose biomass, which assumed to replace corn as the preferred ethanol feedstock over the longer term, would more than offset this advantage when energy use over the full fuel cycle is considered.

In contrast, improvements in fuel processing efficiency are projected to lower upstream energy demands for LPG sufficiently to make its energy use advantage over gasoline considerably larger (about 17%) when both fuel production and use are considered. CNG, which is assumed to be used in dedicated vehicles powered by internal combustion engines, would reduce energy use in vehicle operation slightly (by about 4%) compared to gasoline, but more significantly (about 8%) over the full fuel cycle.

Table 6 also shows that petroleum-based and bio-diesel fuels are each projected to reduce energy use for operating the mix of automobiles and light trucks assumed to be in service during 2025 by about one-third compared to gasoline. Because petroleum diesel (assumed to represent

Table 6. Energy efficiency of gasoline and long-term alternative fuels for light-duty vehicles.

Fuel	Feedstock	Engine Technology (1)	Energy Efficiency (Btu/vehicle-mile)		% change vs. Gasoline	
			Vehicle Operation	Full Fuel Cycle	Vehicle Operation	Full Fuel Cycle
Gasoline	Petroleum	Spark-Ignition ICE	5,651	7,144	--	--
Ethanol (E90)	Biomass (90%)/ Petroleum (10%)	Spark-Ignition ICE; Dedicated Vehicle	5,181	12,724	-8%	78%
CNG	Natural Gas	Spark-Ignition ICE; Dedicated Vehicle	5,430	6,560	-4%	-8%
LPG	Petroleum or Natural Gas	Spark-Ignition ICE; Dedicated Vehicle	5,181	5,915	-8%	-17%
Diesel	Petroleum	Compression-Ignition ICE	3,767	4,606	-33%	-36%
Bio-Diesel (B20)	Petroleum (80%)/ Soy (20%)	Compression-Ignition ICE	3,767	4,865	-33%	-32%
Electricity	Projected U.S. Generating Mix	Battery-Powered Electric Motor	0	4,445	-100%	-38%
Hydrogen (2)	Natural Gas	Spark-Ignition ICE; Dedicated Vehicle	4,236	8,257	-9%	40%

(1) ICE indicates internal combustion engine.

(2) Estimated from limited test data for mid-size automobiles only reported in Szwabowski et al., "Ford Hydrogen Engine Powered P2000 Vehicle," SAE Technical Paper 2002-01-0243, 2002. Percent differences in energy efficiency expressed relative to gasoline-powered automobile.

Source: Estimated using *GREET 1.5a -- Transportation Fuel-Cycle Model*, Argonne National Laboratories, January 2000, using assumptions reported in Table 1. See Appendix Table A-4 for energy consumption rates of individual light-duty vehicle classes.

80% of bio-diesel) also entails lower upstream energy use for feedstock production, fuel refining, and distribution, the overall energy efficiency advantage of these fuels over gasoline is also projected to remain large when measured on a full fuel cycle basis.

As in the near-term analysis, battery-powered electric vehicles nominally eliminate energy use in vehicle operations (because they operate entirely on energy stored in their on-board batteries), but energy losses in generating and transmitting electricity reduce this advantage somewhat when it is measured over the full fuel cycle. Despite this, anticipated improvements in generating efficiency and battery technology suggest that future electric vehicles could improve efficiency significantly (by nearly 40%) over the complete fuel cycle, compared to internal combustion engines using gasoline.

Finally, Table 6 reports that using hydrogen as a fuel could improve the energy efficiency of automobiles powered by internal combustion engines by about 9% when compared to gasoline. Unlike those for the other alternative fuels shown in the table, this result (which was not obtained from the GREET model) is based on very limited test data for prototype mid-size automobiles only, and should be interpreted very cautiously. Assuming that hydrogen fuel is produced from natural gas at decentralized fueling stations (in order to avoid the potentially enormous cost for establishing a hydrogen distribution infrastructure), however, the large energy demand for producing hydrogen would place it at a significant energy efficiency disadvantage compared to gasoline when measured over the complete fuel production and use cycle.

GHG Emissions Rates for Alternative Fuels

Because CO_2 is by far the dominant greenhouse gas produced by combustion of carbon-based fuels, differences in GHG emissions rates for alternative fuels compared to gasoline generally reflect these differences in their energy efficiency when used as fuels for internal combustion engines. However, they also reflect variation in the carbon content of the various alternative fuels, emissions of non-CO_2 greenhouse gases (methane and nitric oxide) generated by combustion of certain fuels, and carbon sequestration "credits" for fuels produced from biomass feedstocks . As Table 7 shows, each of the fuels considered to be a viable longer-term substitute for gasoline would reduce GHG emissions per mile of vehicle travel – some of them very significantly – in both vehicle operation and over the complete fuel production and use cycle.[28]

For example, ethanol is projected to have sharply lower CO_2 and total GHG emissions per vehicle-mile during vehicle operation than gasoline when used in future internal combustion engines (although this result again reflects GREET's convention of crediting carbon sequestration during the growth of ethanol feedstocks against CO_2 emissions from operating vehicles on ethanol). Compressed natural gas and, to a lesser extent, LPG are also projected to lower GHG emissions per vehicle-mile for operating future internal combustion vehicles compared to gasoline.

Petroleum-based diesel and bio-diesel used in compression-ignited engines are anticipated to reduce GHG emissions per mile for operating light-duty vehicles very significantly, although still not to the extremely low levels associated with ethanol. As with ethanol, carbon sequestration during the growth of its biomass feedstock offsets CO_2 emissions from bio-diesel combustion, but this effect is limited because only 20% of bio-diesel is assumed to be derived from biomass.

As in the near-term analysis, battery-powered electric vehicles generate no GHG emissions during vehicle operation, since GREET treats all GHG emissions generated in producing energy to be stored in their on-board batteries as "upstream" emissions. Since hydrogen does not derive energy from stored carbon, its combustion generates only the minimal carbon emissions associated with the presence of lubricating oil in the engine's combustion chambers. As Table 7 shows, this virtually eliminates GHG emissions during vehicle operation using hydrogen as a fuel, although catalytic aftertreatment of exhaust gases might introduce low levels of nitrous oxide (N_2O) emissions. (Unlike those for other fuels included in Table 7, the limited test data available for light-duty internal combustion engines operating on hydrogen do not reflect catalytic aftertreatment of exhaust gases to reduce NOx emissions.)

When GHG emissions in fuel production and distribution are added to those from vehicle operation, Table 7 shows that the same alternative fuels generally offer the largest reductions compared to gasoline-fueled internal combustion vehicles. Compared to gasoline, ethanol is projected to offer dramatic reductions (almost 80%) in full fuel cycle emissions of internal combustion-powered light-duty vehicles, while electricity in battery-powered vehicles is projected to result in significantly lower (nearly 45%) fuel cycle emissions than their gasoline-powered counterparts. Both of these results, however, depend on major innovations: commercialization of technology

Table 7. Greenhouse gas emissions rates for light-duty vehicles using gasoline and long-term alternative fuel/engine technology combinations.

Fuel	GHG Emissions Rates (grams/vehicle-mile)							
	Vehicle Operation				Full Fuel Cycle			
	CO_2 (1)	CH_4 (2)	N_2O (3)	Total (4)	CO_2 (1)	CH_4 (2)	N_2O (3)	Total (4)
Gasoline	396	0.069	0.032	407	514	0.849	0.050	548
Ethanol (E90)	60	0.103	0.032	72	34	0.318	0.240	115
CNG	325	0.343	0.016	337	401	1.510	0.017	438
LPG	371	0.075	0.032	382	426	0.631	0.032	449
Diesel	304	0.013	0.021	311	367	0.413	0.022	383
Bio-Diesel (B20)	249	0.013	0.021	256	367	0.413	0.022	383
Electricity	0	0.000	0.000	0	297	0.444	0.004	308
Hydrogen	1.4	0.000	0.000	1.4	412	0.939	0.003	413

(1) Carbon dioxide; used as numeraire to express "Global Warming Potential" (GWP) of other GHGs.
(2) Methane; GWP relative to carbon dioxide = 21 (100-year value).
(3) Nitrous oxide; GWP relative to carbon dioxide = 310 (100-year value).
(4) CO2 equivalent, with methane and nitrous oxide weighted by GWPs.

Source: Estimated using *GREET 1.5a -- Transportation Fuel-Cycle Model*, Argonne National Laboratories, January 2000, using assumptions reported in Table 1. See Appendix Table A-5 for greenhouse gas emission rates of individual light-duty vehicle classes. Hydrogen data for automobiles only.

for deriving ethanol from cellulosic biomass (such as fast-growing plant species or waste wood products), and development of compact, high energy-density storage batteries to power electric vehicles.

As Table 7 also shows, lower energy use in refining diesel fuel compared to gasoline increases the GHG emissions advantage of petroleum-based and bio-diesel fuels over gasoline to approximately 30% when emissions are measured over the complete fuel cycle rather than in vehicle operation alone. Although GHG emissions generated by the significant use of energy for processing hydrogen from natural gas and compressing it for on-board storage reduce hydrogen's advantage over gasoline, it remains significant (about 25%) when measured on a fuel-cycle basis. Of course, this result depends on commercial-scale development of hydrogen production and retail fueling facilities to support its widespread use as a vehicle fuel within the study's 2025 time horizon. Finally, CNG and LPG are projected by GREET to offer more modest -- but nevertheless important (18-20%) -- reductions in fuel-cycle GHG emissions compared to gasoline when used to power the mix of automobiles and light trucks expected to comprise the U.S. light-duty fleet by that date.

Potential Reductions in Total GHG Emissions

Assuming that internal combustion engine technology continues to dominate the light-duty vehicle fleet over this long-term horizon, the sharply lower GHG emissions rates for some alternative fuels shown in Table 7 would produce significant reductions in total GHG emissions if they could replace the 25% share of gasoline assumed to be feasible within that time frame. Most notably, Table 8 shows that commercializing the processing technology to produce ethanol from cellulosic biomass on a sufficiently large scale to substitute for 25% of projected year-2025 gasoline use could reduce total tailpipe and fuel cycle GHG emissions from the light-duty vehicle fleet by more than 20% from the level of emissions expected to result from forecast travel growth and continued near-exclusive reliance on gasoline.

Using compressed natural gas or LPG to replace one-quarter of total gasoline use by the mix of automobiles and light trucks projected for that date could reduce their total GHG emissions by 2-5%, depending on whether emissions are measured on a "tailpipe" or fuel cycle basis. Replacing the same fraction of gasoline with petroleum-based or bio-diesel fuels could raise this figure to the 6-9% range for GHG emissions from vehicle operations, and to 8% when emissions are measured on the more inclusive fuel-cycle basis.

Electric vehicles appear to have the potential to reduce total fuel-cycle GHG emissions from light-duty vehicle travel by about 11%, if battery technology advanced sufficiently within the long-term horizon of this study to allow manufacturers to offer electric-drive systems across a broad enough range of models -- including automobiles as well as various sizes of light trucks -- to account for 25% of travel by all classes of light-duty vehicles. If instead, advances in battery technology only permitted automobiles (and not light trucks) to achieve the range and performance characteristics necessary to make them attractive alternatives to gasoline vehicles, the reduction in fuel-cycle GHG emissions from using electric automobiles for one-quarter of all light-duty vehicle travel would be about 9% (this figure is not shown in Table 8).

Finally, Table 8 reports that replacing gasoline with hydrogen as a fuel for 25% of travel by all light-duty vehicle classes would reduce GHG emissions over the complete fuel production and use cycle by only about 6%, despite hydrogen's nearly-complete elimination of GHG emissions during vehicle operation. These results illustrate the critical role of energy use and GHG emissions that occur during fuel production and distribution in determining the potential reduction in overall emissions from using alternative fuels.

Although these reductions would represent important contributions to reducing future GHG emissions from transportation and all sources, it is important to keep in mind that achieving the larger reductions would depend on significant innovations in fuel production or vehicle technology. Specifically, the very large reductions estimated to be possible from replacing 25% of gasoline use by ethanol depend on widespread commercialization of technology to produce ethanol from cellulosic biomass. Achieving the more modest GHG emissions reductions possible from replacing that same share of gasoline with diesel fuels or electricity would also require significant improvements in the performance and durability of compression-ignition engines suited for light-duty vehicles, or in battery storage capabilities and electric drive systems. Finally, even the comparatively modest emissions reductions possible from widespread substitution of CNG, LPG, or hydrogen for gasoline fuels would require large-scale investments in new fuel distribution infrastructure, since (unlike ethanol/gasoline blends and diesel fuels) it is unlikely that current gasoline distribution and retailing facilities could be adapted for use with these fuels.

Criteria Pollutant Emissions

Table 9 shows estimates of criteria pollutant emission rates for the projected year-2025 mix of light duty vehicles when operating on gasoline and various alternative fuels.[29] As these figures indicate, most alternative fuels assumed to represent feasible replacements for gasoline over the long-term horizon are projected to result in comparable or even lower tailpipe emissions of criteria pollutants than gasoline. The main source of the sharp reduction in the emission rates shown in Table 9 from the near-term figures shown previously (in Table 5) is anticipated to be continued improvements in

Table 8. Year 2025 greenhouse gas emissions with all-gasoline baseline and 25% gasoline replacement by alternative fuels.

Fuel Mix	Total GHG Emissions (Tg CO_2 equivalent) (1)		Change from Gasoline Baseline (Tg)		% change from Gasoline Baseline	
	Vehicle Operation	Full Fuel Cycle	Vehicle Operation	Full Fuel Cycle	Vehicle Operation	Full Fuel Cycle
100% Gasoline	1,648	2,217	--	--	--	--
25% Ethanol (as E90), 75% Gasoline	1,271	1,731	-377	-487	-23%	-22%
25% CNG, 75% Gasoline	1,577	2,107	-71	-111	-4%	-5%
25% LPG, 75% Gasoline	1,623	2,117	-25	-100	-2%	-5%
25% Diesel, 75% Gasoline	1,551	2,051	-97	-167	-6%	-8%
25% Bio-Diesel (B20), 75% Gasoline	1,495	2,051	-153	-167	-9%	-8%
25% Electricity, 75% Gasoline	1,236	1,975	-412	-243	-25%	-11%
25% Hydrogen, 75% Gasoline	1,238	2,081	-411	-136	-25%	-6%

(1) Teragrams of CO_2 equivalent, computed using 100-year GWPs of 21 for CH_4 and 310 for N_2O. One teragram equals 10^{12} grams, or one million metric tons.

Source: Calculated from Tables 1 and 7.

engine combustion management, together with more effective catalytic aftertreatment of exhaust gases enabled by the reduction in fuel sulfur content required by recently-adopted EPA regulations. These improvements should allow light-duty vehicles using most alternative fuels to comply with the stricter Tier 2 emissions standards scheduled to take effect beginning in model year 2006, which will apply uniformly to all classes of light-duty vehicles (including even the largest light truck models) and regardless of the particular fuel on which they operate.

Nevertheless, the emission rates reported in Table 9 indicate that further reductions of nitrogen oxide (NOx) and particulate matter (PM) emissions for light-duty vehicles operating on petroleum diesel and bio-diesel are likely to be required in order to meet future emission standards. This seems particularly likely if compliance with the pending 8-hour ozone air quality standard proves to require further reductions in motor vehicle NOx emissions, and if accumulating evidence on adverse health effects from human exposure to fine particulate matter prompts the adoption of even more restrictive emission standards for particulate matter within the longer-term time frame.

Thus the technological potential for improving operation and exhaust aftertreatment of compression-ignition engines sufficiently to enable them to comply with stricter emission standards for these pollutants remains an important concern for the ability of diesel fuels to offer viable alternatives to gasoline for reducing light-duty vehicles' GHG emissions. While diesel engine technology is well-

Table 9. Criteria pollutant emissions for gasoline and long-term alternative fuel/engine technology combinations.

Fuel (1)	Engine Technology	Emission Rates (grams/vehicle-mile)				
		VOC (1)	CO (2)	NO_x (3)	SO_x (4)	PM10 (5)
Gasoline	Spark-Ignition ICE	0.130	3.143	0.050	0.008	0.032
Ethanol (E90)	Spark-Ignition ICE; Dedicated Vehicle	0.130	3.143	0.050	0.001	0.028
CNG	Spark-Ignition ICE; Dedicated Vehicle	0.061	2.040	0.050	0.002	0.023
LPG	Spark-Ignition ICE; Dedicated Vehicle	0.068	2.040	0.050	0.000	0.023
Diesel	Compression-Ignition ICE	0.070	4.221	0.107	0.010	0.036
Bio-Diesel (B20)	Compression-Ignition ICE	0.070	4.221	0.107	0.008	0.035
Electricity	Battery-Powered Electric Motor	0.000	0.000	0.000	0.000	0.021
Hydrogen	Spark-Ignition ICE; Dedicated Vehicle	0.008	0.008	0.74 (6)	0.000	0.021

(1) Volatile organic compounds.
(2) Carbon monoxide.
(3) Nitrogen oxides.
(4) Sulfur oxides.
(5) Particulate matter less than 10 microns in diameter.
(6) Engine-out emissions; no exhaust aftertreatment.

Source: Estimated using *GREET 1.5a -- Transportation Fuel-Cycle Model*, Argonne National Laboratories, January 2000, using assumptions reported in Table 1. See Appendix Table A-6 for criteria pollutant emission rates of individual light-duty vehicle classes.

developed and light-duty diesel engines are currently in widespread use, particularly outside the U.S., professional judgments currently differ about the likelihood of achieving the further reductions in their emissions levels and more effective exhaust aftertreatment that will be required to enable light-duty diesel vehicles to meet pending U.S. emission standards.

The Role of Advanced Vehicle Technology

Greenhouse gas emissions from light-duty vehicle use could also be reduced significantly within the longer-term time horizon of this study by advances in engine and vehicle technology, even if gasoline remained the near-exclusive fuel. Still larger reductions in GHG emissions would be possible if the future light-duty fleet included a significant fraction of advanced-technology vehicles operating on alternative fuels. Three advanced engine and drive system technologies that are thought to have the potential for widespread application within this study's long-term time horizon: direct injection internal combustion engines, hybrid internal combustion/electric drive systems, and fuel cell vehicles.

Direct-injection ICEs, which may be either spark- or compression-ignited (diesel), utilize a combustion chamber design that omits the pre-ignition chamber used in conventional internal combustion engines. Hybrid electric vehicles combine a small internal-combustion engine with a storage battery and electric motor, although hybrid designs differ in whether they utilize the power output from their internal combustion engines to power the vehicle directly or to generate electricity to power their electric motors. Fuel cell vehicles would generate electricity using a proton exchange membrane (PEM) fuel cell to power an electric motor, with the fuel cell itself using hydrogen supplied by on-board "reforming" of a hydrocarbon fuel (such as gasoline), or produced at centralized facilities or local fueling stations and stored aboard vehicles.

Table 10 compares the energy efficiency and GHG emissions for light-duty vehicles using these engine technologies and operating on various fuels to those for the improved spark-ignition gasoline engines assumed to represent the "baseline" light-duty engine technology in use as of that date.[30] It also reports estimated reductions in total GHG emissions generated by light-duty travel assuming that each of the engine technology and fuel combinations it includes accounted for the same 25% share of light-duty vehicle travel previously assumed to be possible for alternative fuels.[31]

As Table 10 indicates, direct-injection internal combustion engines, electric hybrid drive systems, and fuel cells offer the potential to improve energy efficiency in vehicle operation by nearly 20% to slightly more than 50% *even if the vehicles using these advanced technologies operated on gasoline.*[32] This would result in comparably large reductions in tailpipe greenhouse gas emissions for these vehicles, even without any displacement of gasoline by alternative fuels. Because GHG emissions in fuel production and distribution are independent of the engine technology in which fuel is used, emissions over the full fuel cycle would also be reduced by comparable proportions – 20% for direct injection ICEs, 46% for electric hybrid systems, and 51% for fuel cells vehicles, as Table 10 reports – even if vehicles using these technologies continued to operate on gasoline.

As Table 10 also shows, total GHG emissions generated by light-duty vehicle use would be reduced by 5% over the full fuel cycle if direct-injection internal combustion engines using gasoline replaced conventional gasoline-powered ICEs for the same 25% of travel previously assumed to be possible for gasoline replacement by alternative fuels. With the same 25% replacement of conventional gasoline ICE vehicles by those using gasoline-electric hybrid drive systems or fuel cell vehicles operating on gasoline, the reductions in total GHG emissions from light-duty vehicle travel would rise to 12-13%, as the table reports.

Comparing these potential reductions in GHG emissions to those for alternative fuels shown previously (in Table 8) suggests that advanced engine technologies could produce emissions reductions larger than almost any alternative fuel (ethanol is the exception) at the *same* level of substitution for conventional internal combustion vehicles (25%) assumed to be possible for gasoline replacement over the longer term. This comparison illustrates that the role of improving vehicle energy efficiency through advanced engine technologies in reducing GHG emissions is critical as long as carbon-based fuels produced from nonrenewable feedstocks remain the dominant form of energy used by transportation vehicles.

Combining Alternative Fuels and Advanced Engine Technologies

Table 10 also provides estimates of the energy efficiency and GHG emissions for vehicles using these advanced engine technologies and operating on selected alternative fuels. Focusing on vehicle operation, direct-injection internal

Table 10. Energy efficiency and GHG emissions from alternative fuels and advanced engine technologies (% changes relative to gasoline in conventional ICE).

Engine Technology (1)	Fuel	Energy Efficiency (Btu/vehicle-mile)		GHG Emissions per Vehicle-Mile		2025 Total GHG Emissions with 25% Use	
		Vehicle Operation	Full Fuel Cycle	Vehicle Operation	Full Fuel Cycle	Vehicle Operation	Full Fuel Cycle
Conventional ICE	Gasoline	--	--	--	--	--	--
Direct Injection ICE	Gasoline	20%	20%	-19%	-20%	-5%	-5%
	Diesel	33%	36%	-24%	-30%	-6%	-8%
	Ethanol (E90)	20%	-55%	-84%	-81%	-21%	-20%
Electric Hybrid	Gasoline	47%	47%	-46%	-46%	-12%	-12%
	Diesel	57%	58%	-50%	-54%	-12%	-13%
	Ethanol (E90)	47%	-2%	-88%	-87%	-22%	-22%
Fuel Cell	Gasoline	50%	50%	-51%	-51%	-13%	-13%
	Methanol	57%	44%	-56%	-55%	-14%	-14%
	Hydrogen	67%	49%	-100%	-58%	-25%	-14%

(1) ICE indicates internal combustion engine.

Source: Estimated using *GREET 1.5a -- Transportation Fuel-Cycle Model*, Argonne National Laboratories, January 2000, using assumptions reported in Table 1. See Appendix Table A-4 for GHG emission rates for individual light-duty vehicle classes.

combustion engines operating on diesel would produce even greater improvements in energy efficiency (33% rather than 20%) and reductions in GHG emissions (24% versus 19%) compared to conventional gasoline-powered ICEs than would direct injection engines operating on gasoline.[33]

Similarly, electric hybrid vehicles that utilized compression-ignited ICEs operating on diesel rather than gasoline-powered engines in conjunction with their battery-powered electric motors could improve the already substantial energy efficiency advantage of hybrid vehicles over conventional gasoline ICEs (from 47% to 57%), while further increasing their advantage over conventional gasoline ICEs in "tailpipe" GHG emissions (to 50%, from the 46% estimated for gasoline-electric hybrid vehicles). As Table 10 also shows, the estimated reductions in fuel cycle GHG emissions for vehicles using diesel direct injection engines (30% compared to conventional gasoline ICEs) or electric hybrid drive systems incorporating diesel ICEs (54%) would be even larger than those during vehicle operation, because of lower GHG emissions in refining diesel compared to gasoline.

Table 10 also reports that operating vehicles powered by direct injection ICEs or hybrid electric drive systems on ethanol could produce still greater reductions in per-vehicle mile and total GHG emissions. As it shows, ethanol-powered direct injection ICEs or hybrid electric vehicles would each reduce "tailpipe" GHG emissions per mile-mile by more than 80% compared to conventional gasoline ICEs, although this figure would be reduced slightly when measured over the full fuel cycle.

Similarly, fuel-cell powered vehicles using either methanol (to produce hydrogen in on-board reformers) or hydrogen produced at fueling stations from natural gas could reduce GHG emissions measured over the complete fuel cycle by well over 50% compared to those for light-duty vehicles using conventional gasoline-powered ICEs. Still larger reductions in fuel-cycle GHG emissions from fuel-cell vehicles would be possible if hydrogen could be produced through hydrolysis powered by low-emissions sources of electrical energy such as nuclear or solar power. However, prospects for commercial development of this technology and for construction of the extensive hydrogen distribution and retailing infrastructure it would require remain highly uncertain within even the longer-term time horizon of this study.

As Table 10 also reports, a future light-duty fleet that incorporated 25% of vehicles powered by direct injection engines could reduce total fuel-cycle GHG emissions from light-duty vehicles by 8% if these vehicles were powered by diesel, and by as much as 20% if they operated on cellulosic ethanol. If the same fraction (25%) of light-duty vehicles were instead powered by hybrid electric drive systems, the corresponding reductions in fuel-cycle GHG emissions would reach 13% if diesel fuel were used, and as much as 22% if these vehicles used ethanol as a fuel. Finally, as Table 10 shows, replacing one-quarter of conventional gasoline powered light-duty vehicles with fuel cell vehicles powered by hydrogen (manufactured by either on-board reforming of methanol or processing of natural gas at fueling stations) could reduce fleet-wide GHG emissions by 14% on a fuel-cycle basis. Again, if hydrogen could be produced through electrolysis powered by low-GHG emission sources of electricity, this figure would rise significantly, but commercialization of this technology and installation of the accompanying distribution infrastructure appear to be more distant prospects.

These comparisons suggests that combining the potential improvement in energy efficiency from advanced-technology vehicles with the substitution of certain alternative fuels for a significant fraction of gasoline could significantly reduce GHG emissions from light-duty vehicle use. Of course, this strategy would entail significant costs, both for encouraging widespread production and purchases of advanced-technology vehicles and for promoting the substitution of alternative fuels for a significant fraction of gasoline. And while such a strategy might not by itself allow the ambitious GHG-reduction targets based on 1990 emissions levels specified in international agreements to be met, it could offset much of the growth in emissions expected to result from continued increases in light-duty vehicle travel with continued dominance of conventional internal combustion engine technology and near-exclusive reliance on gasoline.

Cost-Effectiveness of Alternative Fuels

Because a wide variety of transportation, industrial, and other activities generate emissions of greenhouse gases, there are many potentially effective strategies for reducing those emissions. Developing policies to meet GHG reduction targets while minimizing the resulting impact on U.S. economy activity requires a careful assessment and comparison of the costs of reducing GHG emissions through various strategies. In an effort to contribute to such an assessment, this section estimates costs for producing and distributing alternative transportation fuels, and for adapting light-duty vehicles to operate on those fuels instead of gasoline.

It uses these cost estimates in conjunction with the estimates of GHG emissions reductions reported previously to calculate the cost-effectiveness – or cost per unit of emissions eliminated -- of shifting a fraction of light-duty vehicle use from gasoline to alternative fuels as a strategy for reducing U.S. GHG emissions. Because of the extreme uncertainty inherent in long-term forecasts of vehicle production costs and fuel prices, this study analyzes the cost-effectiveness of alternatives to gasoline for reducing GHG emissions only within the 10-year near term horizon.

Major Elements of Alternative Fuel Costs

Substituting alternative fuels for gasoline and diesel entails three main categories of additional costs. First, producing vehicles with the capability to operate on alternative fuels entails higher costs than manufacturing gasoline-powered vehicles. Equipping or modifying conventional engine designs to operate on fuels other than gasoline or diesel typically entails additional design, tooling, and production costs. Although these costs are generally minor for fuels with properties similar to gasoline, they can be significant if entirely different engine designs must be substituted for conventional spark-ignited internal combustion engines, or if vehicles' fuel storage and delivery systems require extensive modification to accommodate alternative fuels. For example, adapting light-duty vehicles designed to operate on gasoline to use diesel instead requires substituting compression-ignition engines, which are considerably more costly to produce, for conventional spark-ignited internal combustion engines.

Costly modifications to vehicles' on-board fuel storage tanks and fuel delivery systems are also required to permit their operation on certain alternative fuels, such as CNG. An extreme example is bi-fuel vehicles, which require separate fuel storage and delivery systems to permit their operation on two different fuels, usually gasoline and CNG. Designing light-duty vehicles to operate on fuels with lower energy densities than gasoline also requires fuel storage tanks to be larger in order to provide an equivalent mileage range between required refueling stops. If manufacturers elect to provide equivalent range in alternative fuel vehicles, production costs will increase slightly, while if not, owners will incur added costs for time and inconvenience associated with more frequent refueling.[34]

Not surprisingly, costs for engine re-design and modifications to fuel delivery systems are considerably lower when vehicles are designed and manufactured to operate on alternative fuels than when gasoline vehicles are subsequently modified to use alternative fuels. This analysis assumes all that vehicles operating on alternative fuels are designed and initially manufactured to use those fuels, rather than subsequently retrofitted or converted to use gasoline substitutes. It also assumes that vehicles operating on each alternative fuel would be produced in sufficient volumes to fully exploit any scale economies in the production and assembly of their specialized fuel storage and delivery systems, so that vehicles operating on these fuels are manufactured at costs as closely comparable to gasoline vehicles as current technology permits.

Expanding the use of most alternative fuels also requires increased capital investments in facilities for producing or extracting the feedstocks from which they are derived, for refining alternative fuels from these feedstocks, and for distributing them to retail stations or other fueling facilities. These investments are *in addition* to those that would be required to support expected growth in the use of these fuels for current non-transportation applications. Thus for example, expanding the use of natural gas as a transportation fuel would require investments in expanded fuel production facilities and pipelines over and above those already planned to allow its increased use for electricity generation and residential heating, as well as extensive investments to modify retail fueling stations to accommodate natural gas.

Finally, alternative fuels themselves may be more or less expensive for vehicles to use than gasoline or diesel when measured on a cost per vehicle-mile basis. This can occur either because the equivalent prices per unit of energy contained by alternative fuels differ from conventional fuels, or because the efficiency with which this stored energy can be converted to propulsion energy differs from conventional fuels. Natural gas and diesel-type fuels are typically less expensive per unit of energy to produce than gasoline, whereas alcohol fuels (ethanol and methanol), liquefied petroleum gas, and electricity are often (but not always) more costly to produce per unit of energy content.

As the analysis of energy use and GHG emissions in the previous section showed, the energy conversion efficiency of current light-duty engine technologies operating on some of these fuels is slightly lower than that for gasoline, while for others (notably diesel fuels) it is considerably higher. Thus on balance, alternative fuels can be more or less costly per vehicle-mile of travel than gasoline, and can either increase or reduce the total costs for using alternative fuels as a strategy to reduce GHG emissions.

Costs for Near-Term Alternative Fuels

Estimates of increased production costs for modifying light-duty vehicles' engines and on-board fuel delivery systems to operate on the alternative fuels assumed to be widely available within the near term (year 2010) were obtained from published research. This analysis also relies upon previously reported estimates of costs for constructing new or expanded infrastructure required for the production, storage, and distribution of different fuels. Vehicle production and infrastructure cost estimates were assembled from their original sources and updated to current dollars to provide comparability, and the estimates of fuel infrastructure costs were also adjusted to a per-vehicle basis using current relationships among the number of vehicles operating on each fuel, total consumption of that fuel, and the number of retail fueling stations offering it for sale.[35]

Vehicle Production Costs

As Table 11 indicates, costs for modifying light-duty vehicles to permit their operation on ethanol as well as on gasoline (thus the term flexible-fuel vehicle) are slightly less than $300 per vehicle, since only minor modifications to gasoline vehicles' fuel delivery systems are required to permit their operation on ethanol. In contrast, strengthening vehicles' fuel tanks and modifying their fuel delivery systems to allow exclusive operation on compressed natural gas raises this figure considerably, to more than $1,900 per vehicle. Retaining CNG vehicles' capability to operate on gasoline as well raises incremental production costs further to slightly over $2,100 per vehicle, since these "bi-fuel" vehicles require fully independent gasoline and CNG storage and delivery systems. Incremental production costs for dedicated LPG vehicles fall between those for ethanol and CNG – nearly $750 per vehicle, as Table 11 reports – since their fuel delivery systems are similar to those for CNG, but LPG utilizes low-pressure storage tanks that more closely resemble conventional gasoline tanks.

Replacing light-duty gasoline engines with diesel engines offering equal performance, durability, and capability for complying with the pending Tier 2 emission standards has been estimated to entail added manufacturing costs in the range of $2,500 per vehicle, as Table 11 reports. Finally, the current manufacturing cost premium for battery-powered vehicles over their gasoline counterparts at large production volumes appears to be slightly more than $4,000 per vehicle. Since electric motors are a well-developed technology, most of this increased investment represents costs for the large on-board storage batteries that electric vehicles currently require. Even at this additional cost, current battery technology limits the performance and range of electric vehicles to levels well below those offered by gasoline vehicles, thus limiting their substitutability for gasoline vehicles for many household trips.

Fuel Infrastructure Costs

Additional capital costs for fuel delivery infrastructure were converted from their original terms to a per-vehicle basis in order to facilitate their comparison to one another, as well as to estimated costs for required vehicle modifications. Fuel infrastructure costs also vary widely among the alternative fuels included in this study. At one extreme, the ready adaptability of much of gasoline delivery infrastructure to use with other liquid fuels minimizes the incremental infrastructure costs

Table 11. Estimated capital costs for alternative fuel vehicle production and infrastructure provision.

Fuel	Engine Technology	Incremental Capital Cost per Vehicle (2000 $) (1)			Total Incremental Capital Costs (2000 $, billions)		
		Vehicle Production	Fuel Infra-structure (2)	Total	Vehicle Production (3)	Fuel Infra-structure (3)	Total
Ethanol (E85)	Spark-Ignition ICE; Flexible-Fuel Vehicle	$284	$128	$412	$14.1	$6.4	$20.5
CNG	Spark-Ignition ICE: Bi-Fuel Vehicle	$2,104	$398	$2,502	$104.5	$19.8	$124.3
CNG	Spark-Ignition ICE: Dedicated Vehicle	$1,914	$398	$2,313	$47.6	$9.9	$57.5
LPG	Spark-Ignition ICE: Dedicated Vehicle	$741	$222	$963	$18.4	$5.5	$23.9
Diesel	Compression-Ignition ICE	$2,500	$34	$2,534	$62.1	$0.8	$63.0
Bio-Diesel (B20)	Compression-Ignition ICE	$2,500	$34	$2,534	$62.1	$0.9	$63.0
Electricity	Battery-Powered Electric Motor	$4,100	$120	$4,220	$101.9	$3.0	$104.8

(1) Source: see Appendix Table A-7 for original estimates and derivation of values.

(2) Includes estimated investments for necessary storage, distribution, and retailing infrastructure. Source: calculated from values reported in Appendix Table A-7 using procedure described in text.

(3) Assumes 10% of projected year-2010 light-duty vehicle fleet (27 million vehicles) must be capable of operating on alternative fuel if dedicated vehicles are used, and 20% (54 million vehicles) if flexible-fuel or bi-fuel vehicles are used.

for these fuels. As Table 11 shows, this is particularly true for petroleum diesel and bio-diesel, which entail additional fuel infrastructure costs compared to gasoline of less than $35 per vehicle.

Although necessary modifications for handling alcohol fuels such as ethanol raise their fuel infrastructure costs from the minimal levels required for diesel fuels, they nevertheless remain relatively modest -- about $130 more per vehicle than gasoline distribution facilities -- by comparison to those for most other fuels. Costs for electric vehicles, which are limited to those for installation of medium-voltage charging facilities in households and at some public or retail facilities, are also estimated to be similarly modest (about $120 more per vehicle than gasoline), as Table 11 also shows.

However, constructing the pressurized fuel storage and distribution systems required to deliver compressed natural gas or liquefied petroleum gas to retail stations raises the incremental infrastructure costs for these fuels significantly. As Table 11 reports, additional investments in the infrastructure required for CNG storage, distribution, and retail vehicle fueling amount to almost $400 more per vehicle than those for gasoline. Similarly, the specialized capital facilities required to distribute, store, and refuel light-duty vehicles with LPG on a widespread basis amount to over $220 per vehicle, as Table 11 also shows.

Total Capital Investments

By expressing these incremental capital costs on a per-vehicle basis, they can be used to estimate the *total* additional costs for vehicle production and expanded fuel infrastructure that would be necessary to allow each alternative fuel to displace any assumed fraction of projected gasoline use. These costs depend on the per-vehicle values for incremental vehicle production and fuel infrastructure costs reported in Table 11, the total number of light-duty vehicles projected to be in service, and the fraction of gasoline use replaced by each alternative fuel, assumed to be 10% by the study's 2010 near-term horizon. For fuels that would be used in dedicated vehicles, that same fraction of the vehicle fleet would be required to be alternative fuel-capable.

In contrast, flexible-fuel and bi-fuel vehicles -- those capable of operating on gasoline as well as on an alternative fuel -- are assumed to operate only half the time on an alternative fuel, so that 20% of the total fleet would be required to consist of these vehicles. (This assumption appears to be extremely generous by comparison to current experience with flexible-fuel and bi-fuel vehicles, although operation of those vehicles on alternative fuels is currently restricted by the limited number of retail outlets for non-gasoline fuels.) The total size of the U.S. light-duty vehicle fleet during 2010 was estimated from the forecast of total vehicle-miles for the near-term horizon year and average vehicle utilization, which was assumed to increase slightly from its current figure of approximately 12,000 miles per vehicle per year.

The total additional capital investment in vehicles and fuel infrastructure necessary to replace 10% of projected gasoline use with ethanol appears to be relatively modest – about $21 billion during the year 2010, as Table 11 shows. Estimated additional vehicle and fuel infrastructure costs for replacing the same fraction of gasoline with LPG would be slightly larger, while those for replacing 10% of gasoline with CNG would approach $60 billion if dedicated vehicles were employed, and nearly double that figure with bi-fuel CNG vehicles.

Primarily because of the significant cost premiums for manufacturing diesel engines and exhaust aftertreatment systems that could allow light-duty vehicles to meet pending federal emission standards, the additional capital investment required to substitute petroleum diesel or bio-diesel for 10% of projected year-2010 gasoline use would be slightly higher than those for dedicated CNG vehicles – about $63 billion, as Table 11 shows. Finally, the incremental capital investment for replacing 10% of gasoline vehicle travel with electric vehicles are estimated to exceed $100 billion, mainly because of the significant additional costs for supplying the batteries used to power them. These figures each represent the *additional* capital investments in vehicles and fuel distribution facilities beyond those necessary to produce additional gasoline-powered vehicles and expand the nation's current gasoline distribution system sufficiently to meet increased demand for light-duty vehicle travel by 2010.

Fuel Supply and Total Costs

By converting these total capital costs for producing alternative-fuel vehicles and expanding fuel distribution infrastructure to their annual equivalent values, which are reported in Table 12, the portion of these costs that are attributable to achieving the previously-estimated GHG reductions during the near-term horizon (2010) year can be determined.[36] Adding the incremental costs for supplying the volume of each fuel necessary to replace 10% of projected gasoline use during that year, also shown in Table 12, yields an estimate of the annual cost for achieving the resulting reduction in GHG emissions. For certain alternatives to gasoline this incremental fuel production cost is negative, indicating that these fuels are *less* costly to produce per unit of energy than gasoline, so that substituting them for gasoline results in the cost reduction shown in Table 12.[37]

As an illustration, significant cost savings for refining diesel compared to gasoline partly offset diesel vehicles' higher manufacturing and fuel infrastructure costs, thus making total year-

2010 costs for replacing 10% of gasoline use with petroleum diesel the lowest among the gasoline alternatives considered (slightly over $2 billion). Because its soy-based component is much more costly to produce, replacing a similar fraction of gasoline with bio-diesel would be nearly twice as expensive -- $3.9 billion, as Table 12 shows – as diesel derived entirely from petroleum, despite the fact that its soy component accounts for only one-fifth of the fuel's total volume. Compressed natural gas also features lower fuel production costs per unit of energy than gasoline, so that replacing 10% of projected 2010 gasoline use with CNG used in dedicated vehicles would entail costs comparable to those for bio-diesel (about $4.0 billion). Relying on CNG used in bi-fuel vehicles to replace an equivalent fraction of gasoline use would be much more costly -- more than $11 billion -- because of these vehicles' higher production costs and slightly lower energy efficiency compared to those designed to operate solely on CNG, as well as because many more of these vehicles would be required (since they are assumed to operate on CNG for only half of the mileage they are driven).

In contrast, sharply higher fuel production costs for ethanol more than offset its advantages over other gasoline substitutes from low vehicle production and fuel infrastructure costs. As a consequence, Table 12 shows that total annual costs for displacing 10% of projected gasoline use during 2010 with ethanol would exceed $32 billion.[38] Liquefied petroleum gas also has significantly higher production costs per unit of energy than gasoline, so that total estimated costs for replacing one-tenth of projected gasoline use with LPG exceed $8 billion despite low additional costs for vehicle manufacturing. Finally, as Table 12 reports, sharply higher costs for vehicle production and for electricity generation and distribution make electric vehicles a costly alternative to gasoline, with estimated year-2010 costs to achieve the 10% gasoline replacement target approaching $16 billion.

Alternative Fuel Cost-Effectiveness

Combining the estimates of total annual costs for using each alternative fuel to partly replace gasoline shown in Table 12 with the previous estimates of GHG reductions over the full fuel cycle (which are shown again in Table 12) allows an assessment of each fuel's cost-effectiveness as a strategy for reducing GHG emissions. Specifically, dividing the estimated total cost for replacing 10% of projected gasoline use in the year 2010 with each alternative fuel by the resulting annual reduction in GHG emissions during 2010 yields the figures for total cost per ton of emissions avoided reported in the last column of Table 12. Ranking these values for different fuels allows their cost-effectiveness as strategies for reducing GHG emissions to be assessed, and to be compared to the cost-effectiveness of competing strategies to reduce emissions from transportation or other sources.

As these calculations show, the combination of very low incremental costs for replacing gasoline with petroleum diesel and the moderate reduction in GHG emissions that would result implies a cost of approximately $50 per ton of emissions avoided, by far the lowest cost for any of the fuels analyzed. Substituting a soy-derived component for 20% of petroleum diesel (bio-diesel, or B20) would add significantly to costs and reduce the impact on GHG emissions from substituting that fuel for gasoline, thus raising the cost per ton of emissions removed by bio-diesel to the $100 range. Substituting CNG for 10% of gasoline use would entail slightly higher costs than bio-diesel, but the much smaller emissions reduction that would result would increase costs per ton of GHG emissions avoided to more than $300 if CNG were used in dedicated vehicles, and to nearly $1,400 per ton avoided if bi-fuel vehicles were relied exclusively upon.

Substituting corn-based ethanol for gasoline could achieve emissions reductions even larger than those from using petroleum diesel, as Table 12 shows, but high costs for producing ethanol would increase the cost of eliminating GHG

Table 12. Cost-effectiveness of near-term alternative fuels in reducing GHG emissions from light-duty vehicles.

Fuel	Annualized Incremental Cost for Assumed Alternative Fuel Use (2000 $, billions) (1)				Annual GHG Reduction (Tg CO_2 equivalent) (3)	Cost/Ton of Emissions Avoided (2000 $) (4)
	Vehicle Production (2)	Fuel	Fuel Infra-structure (2)	Total		
Ethanol (E85)	$1.5	$30.0	$0.5	$32.1	50.6	$630
CNG (Bi-Fuel Vehicles)	$11.5	-$1.6	$1.7	$11.5	8.4	$1,380
CNG (Dedicated Vehicles)	$5.2	-$2.0	$0.8	$4.0	13.0	$310
LPG	$2.0	$5.7	$0.5	$8.2	18.9	$430
Diesel	$6.8	-$4.8	$0.1	$2.1	42.7	$50
Bio-Diesel (B20)	$6.8	-$2.9	$0.1	$3.9	38.9	$100
Electricity	$11.2	-$14.4	$0.3	-$3.0	47.0	-$60

(1) Calculated from values in Table 11 and Appendix Table A-7 using procedure described in text.

(2) Calculated using 7% annual discount rate and assumed lifetimes of 15 years for vehicles and 25 years for fuel infrastructure.

(3) Reduction over Full Fuel Cycle; calculated from entries in Table 4. One Teragram (10^{12} grams) equals one million metric tons.

(4) Equals annualized total cost divided by annual reduction in GHG emissions, expressed in metric tons.

emissions to more than $600 per ton. Replacing 10% of gasoline use with LPG would entail lower incremental fuel production costs than ethanol but also a smaller reduction in emissions, thus raising costs per ton of emissions avoided still further, to the $400-450 range. Finally, replacing 10% of travel by gasoline-powered internal combustion vehicles with battery-powered electric vehicles would reduce GHG emissions by nearly as much as would the use of ethanol, while sharply lower costs for electricity generation would partly offset high costs for electric vehicle production, making their use a moderately expensive GHG-control strategy among the fuels considered here, with costs approaching $350 per ton removed.

Improving the Cost-Effectiveness of Alternative Fuels

Although extensive recent research has attempted to estimate vehicle manufacturing, fuel production, and infrastructure provision costs to support wider use of alternative fuels over the longer term, most of these estimates are too approximate or speculative to support reliable analysis of the cost-effectiveness of using alternative fuels to replace gasoline as a longer-term GHG reduction strategy. This is particularly true for fuels that are not now in commercial-scale production (such as ethanol derived from cellulosic biomass) or that would be used in conjunction with engine and drive system technologies that are still under development (for example, hydrogen used to power fuel cell vehicles). Nevertheless, the results of the near-term cost-effectiveness estimates shown in Table 12 do suggest some specific developments that will be required to improve the cost-effectiveness of employing various alternative fuels to reduce GHG emissions over the longer term.

For example, the incremental cost figures shown in Tables 11 and 12 indicate that the key to making ethanol use by light-duty vehicles a cost-effective GHG control strategy is to reduce the currently high cost of producing the fuel itself, since the additional costs for modifying vehicles and gasoline distribution infrastructure to accommodate ethanol are already minimal. The most promising approach for doing so appears to be commercialization of technology to produce ethanol from waste or low-grade cellulosic biomass (such as scrap wood or rapidly growing plant species), since this would lower feedstock costs significantly compared to corn, currently the preferred ethanol feedstock. It may also be possible to reduce or even eliminate incremental vehicle production costs to allow its use by blending ethanol at a lower percentage (compared to the 85% it represents in E85) with gasoline but doing so on a more widespread basis, although additional costs for adapting vehicles to operate on ethanol are not a major source of its current cost disadvantage compared to gasoline.

Increasing the cost-effectiveness of LPG in reducing GHG emissions also seems likely to require lowering the costs of supplying the fuel itself, although this may be difficult to achieve because LPG is largely a by-product of crude petroleum and natural gas extraction that is produced in roughly fixed proportions with those feedstocks. In contrast, improving the cost-effectiveness of CNG as a gasoline substitute will require significant reductions in the costs of modifications to vehicles' fuel storage and delivery systems to accommodate its use. However, the potential improvement in CNG's cost-effectiveness as a GHG control measure is limited by the modest reductions in GHG emissions that would result from its substitution for gasoline, even on a large-scale basis.

As the cost figures in Tables 11 and 12 indicate, the near-term cost-effectiveness disadvantage of electric vehicles stems from both high vehicle manufacturing costs (primarily for electric storage batteries), and high costs of electricity delivered to residential locations. Breakthroughs in battery technology that lower production costs significantly and reductions in delivered electricity costs are both likely to be necessary for replacing gasoline vehicles with electric vehicles to become a cost-effective GHG control strategy over the longer term.

While replacing gasoline with petroleum diesel fuel appears to be a reasonably promising GHG control strategy even in the near term, its cost-effectiveness might approach those of some low-cost strategies for reducing emissions from non-transportation sources if expenses for manufacturing high-performance compression ignition engines suitable for light-duty vehicles could be reduced. To some extent, this might be achieved simply as a result of higher-volume production of light-duty diesel engines, although some efforts to tailor engine designs more carefully to light-duty vehicle performance demands may be necessary as well.

Ensuring the cost-effectiveness of petroleum diesel for reducing GHG emissions is also likely to depend on reducing its sulfur content sufficiently to permit effective aftertreatment of diesel exhaust (to reduce NOx and fine particulate emissions to levels permissible under pending emissions standards), without sacrificing diesel fuel's refining cost advantage over gasoline. Improving the cost-effectiveness of bio-diesel fuel as a GHG control measure will also require reductions in the cost of refining its biomass component, which may entail substitution of other feedstocks for soy oil.

Finally, as the discussion suggested previously, the improved energy efficiency of advanced-technology engines and propulsion systems for light-duty vehicles – direct-injection gasoline or diesel engines, hybrid electric drive systems, and fuel cells – also offers the potential for sizeable reductions in light-duty vehicles' GHG emissions over the longer term. If improvements in their designs continue and commercially viable manufacturing processes can be developed for these advanced technologies, they may be offered in a sufficiently wide range of vehicle models at incremental costs low enough to result in their widespread incorporation into the light-duty vehicle fleet over the longer term. Combining the resulting increase in vehicle efficiency with the substitution of cost-effective alternative fuels for a significant fraction of gasoline could produce large reductions in transportation sector GHG emissions over the longer term at reasonable additional costs.

Potential Concerns with Alternative Fuels

In addition to required investments in new infrastructure and higher costs for producing some alternative fuels, there are likely to be other potential impediments to their widespread substitution for current transportation fuels. These include possible difficulties in extracting or importing increased supplies of the feedstocks necessary to produce some alternative fuels, increasing the production of alternative fuels that are currently available on only a limited scale, and expanding their distribution and retailing networks to duplicate (or at least approach) the ubiquitous availability of conventional fuels.

In addition, the widespread distribution, storage requirements, and increased handling of alternative fuels that would be entailed by their common use poses potential safety, health, and environmental hazards that would need to be carefully addressed in order to achieve public acceptance of these fuels. This section briefly estimates increased production and feedstock demands for producing alternative fuels, outlines their requirements for retail fueling infrastructure, and identifies potential safety and environmental hazards posed by their expanded distribution and use.

Fuel and Feedstock Production Demands

Potential complications in producing and distributing the required volumes of some alternative fuels are one possible impediment to their widespread use as replacements for gasoline. Because total gasoline production is so large, the alternative fuel production demands to meet even the modest gasoline replacement targets assumed for the near term could require major increases in refining or production capacity for most fuels, as well as in storage and distribution infrastructure, and in retail or home fueling facilities. Significant increases in production or imports of the feedstocks used to produce alternative fuels are also likely to be required, except in the case of fuels that could be refined from the same petroleum feedstocks that would otherwise be used to produce the gasoline they replace. This section estimates the increases in alternative fuel and feedstock production that would be required to meet the gasoline replacement targets assumed in this study, and compares them to production volumes expected in the absence of efforts to promote use of alternative fuels.

Near-Term Production Demands

The total volumes of each alternative fuel that would be necessary to achieve the 10% gasoline displacement target for 2010 can be calculated from the total fuel energy required to replace that fraction of the forecast volume of gasoline use. Because both the energy content per unit of volume and the efficiency with which it can be converted to propulsion energy by available engine technologies vary among fuels, the total volume of each fuel that would be necessary to replace the assumed fraction of gasoline also differs. As Table 13 illustrates, however, even the seemingly modest objective of replacing 10% of gasoline use would require very large increases in the production levels for some alternative fuels from the levels that would be expected in the absence of efforts to promote their use as substitutes for gasoline.[39]

For example, 25 billion gallons of fuel ethanol (to be blended with 15% gasoline) would be required to replace 10% of light-duty vehicle gasoline use projected for the year 2010, about 8.5 times the production volume expected without more aggressive efforts to promote its use. The slightly higher energy efficiency of

Table 13. Alternative fuel production volumes necessary to replace 10% of projected 2010 gasoline use.

Fuel	Form	Unit of Measure	Energy Density (Btu/unit) (1)	Fuel Energy (quadrillion Btu) (2)	Fuel Volume (billion units) (3)	% Increase from Baseline Production (4)
Ethanol (E85)	liquid	gallon	81,780	2.04	25.0	850%
CNG	gaseous	cubic foot	928	1.94	2,095	7%
LPG	liquid	gallon	84,000	1.81	21.5	56%
Diesel	liquid	gallon	128,500	1.34	10.4	18%
Bio-Diesel (B20)	liquid	gallon	126,220	1.34	10.6	-- (5)
Electricity	--	kWh	11,765	1.97	167	4%

(1) Source: Wang, Michael, *GREET 1.5a -- Transportation Fuel-Cycle Model: Volume 1, Methodology, Use, and Results*, ANL/ESD-39, Argonne National Laboratories, 1999, Table 3.3 (low heating values).

(2) Calculated from forecast of light-duty vehicle travel reported in Table 2.

(3) Equals total alternative fuel energy use divided by energy density of fuel.

(4) Equals total alternative fuel volume as a percent of production forecast for 2010 reported in U.S. Energy Information Administration, *Annual Energy Outlook 2000 -- Reference Case Forecast*, Appendix A, Tables A-1 and A-2. See Appendix Table A-8 for baseline production forecasts.

(5) No production anticipated in absence of use as a transportation fuel; hence, percent increase cannot be calculated.

light-duty vehicles operating on LPG compared to ethanol and its greater energy density combine to lower the volume of LPG required to reach the same gasoline replacement target to 21.5 billion gallons, although this still represents more than a 50% increase in the expected volume of LPG production during 2010. As Table 13 also shows, replacing 10% of 2010 gasoline use would require more than 10 billion gallons of petroleum diesel or bio-diesel. This would amount to nearly a 20% increase in refining of petroleum diesel despite its already large-scale production for use by heavy-duty vehicles, and would require the development of entirely new capacity to produce the soy component of bio-diesel on a commercial scale.

In contrast, much more modest increases in fuels already in widespread use outside the transportation sector would be required to replace 10% of gasoline use. As Table 13 reports, about 2.1 trillion cubic feet of CNG would be required to meet the same gasoline replacement target, although this would represent only about a 7% increase in forecast 2010 natural gas production. Replacing 10% of gasoline-powered vehicle travel with electric vehicles would require an even smaller percentage increase in the projected future level of electricity generation and transmission (about 4%), as the table also shows.

Correspondingly large increases in production, storage, and distribution capacity would also be necessary for some alternative fuels. At the very least, extensive modification of gasoline retailing infrastructure would be required to accommodate liquid fuels such as ethanol or diesel. For fuels such as natural gas or LPG, constructing entirely new retail distribution facilities or installing fuel storage capacity and refueling equipment at existing gasoline retailing stations would be required. Even with electricity, installation of charging facilities in millions of dwellings, large numbers of publicly-accessible locations, and probably at some large employment sites would be necessary.

In turn, the increases in feedstock production volumes that would be necessary to support large-scale gasoline replacement depend on these estimated fuel volumes and on fuel yields per unit of feedstock, and these are calculated in Table 14. These estimates assume some increases in current feedstock yields for alternative fuels by 2010, as a result of continued gradual improvements in refining or production processes. Increases in feedstock production demands would be generally modest – well under 10%, as Table 14 shows -- if an alternative fuel derived from petroleum or natural gas replaced part of gasoline use, because both feedstocks are already produced or imported on a large scale to meet a variety of final energy use demands (including widespread use of natural gas as a fuel itself).

However, increases in biomass feedstock output necessary to meet ethanol or bio-diesel fuel demands would be very large by comparison to current production of corn and soy, which is primarily for use as animal feeds and in food processing. Even with improved processing yields, for example, more than 8 billion bushels of corn would be required to produce sufficient ethanol to allow an E85 blend to replace 10% of projected year-2010 gasoline use. This would require increasing projected U.S. corn production during that year by more than 75%, which -- even assuming no decline in crop yields accompanying the expansion in land area under cultivation -- would require significant increases in water consumption and fertilizer use in corn farming, as well as in energy use for land cultivation, harvesting, and transportation. Similarly, producing the soy component of bio-diesel even in the limited volume necessary to replace 10% of projected gasoline use would require about 16 billion pounds of additional soybean production during 2010, an increase of more than 70% from forecast harvest levels. Again, this would entail similar increases in land devoted to soybean cultivation, agricultural inputs, and energy use in farming and transportation.

Long-Term Production Demands

Fuel production volumes and associated feedstock demands necessary to support alternative fuel use could be significantly larger over the longer term, particularly if internal-combustion engines remained the dominant technology in use for light-duty vehicles. This is partly because the longer-term target for gasoline replacement is much more ambitious (25% versus 10% in the near term), and partly because the projected volume of gasoline use would be larger as a consequence of travel growth. As Table 15 reports, increases in alternative fuel production volumes necessary to reach 25% displacement of gasoline would generally be three to four times larger than required to meet the near-term gasoline replacement target, and would again be very large by comparison to baseline production volumes for all fuels except CNG and electricity.

The corresponding increases in feedstock production demands would be partly offset by expected improvements in refining and processing yields over the 25-year longer-term horizon, but Table 16 shows that they would be very large for some alternative fuels, particularly those produced from biomass (ethanol and the soy-derived component of bio-diesel).[40] Development of entirely new feedstocks over this period is assumed to occur only for ethanol, where the advent of commercial-scale technology for using cellulosic material would allow herbaceous or woody biomass to replace corn as a feedstock. As in the near-term, widespread use of petroleum and natural gas as feedstocks for conventional fuels (and of natural gas as a fuel itself) means that increased demands for alternative fuels derived from these same feedstocks would translate into relatively modest -- and thus probably manageable -- increases in total petroleum or natural gas demand.

Table 14. Feedstock demands for alternative fuel production volumes necessary to replace 10% of projected 2010 gasoline use.

Fuel	Feedstock	Units of Measure		Fuel Volume (billion units) (1)	Feedstock Yield (fuel units per feedstock unit) (2)	Feedstock Demand (billion units) (3)	% Increase from Baseline Production (4)
		Fuel	Feedstock				
Ethanol (E85)	Corn	gallon	bushel	25.0	2.533	8.38	76%
CNG	Natural Gas	cubic foot	cubic foot	2,095	0.872	2,404	7%
LPG	Petroleum	gallon	gallon	21.5	1.381	15.6	6%
LPG	Natural Gas	gallon	cubic foot	21.5	0.010	2,181	7%
Diesel	Petroleum	gallon	gallon	10.4	0.850	12.3	5%
Bio-Diesel (B20)	Petroleum	gallon	gallon	8.5	0.850	10.0	4%
	Soy Oil	gallon	pound	2.1	0.130	16.3	72%
Electricity	Projected U.S. Mix	kwh	various	167	(5)	(5)	(5)

(1) Source: Table 13.

(2) Estimated from values reported in Wang, Michael, *GREET 1.5a -- Transportation Fuel-Cycle Model: Volume 1, Methodology, Use, and Results*, ANL/ESD-39, Argonne National Laboratories, 1999, Tables 3.3, 4.3, 4.11, 4.25, and 4.28.

(3) Equals alternative fuel volume divided by feedstock yield.

(4) Equals feedstock volume as a percent of production forecast for 2010 reported in U.S. Energy Information Administration, *Annual Energy Outlook 2000 -- Reference Case Forecast*, Appendix A, Tables A-1 and A-2. For biomass feedstocks, equals feedstock volume as a percent of total production for most recent year available, reported in U.S. Department of Commerce, Statistical Abstract of the United States, 1999, Tables 1129 and 1130. See Appendix Table A-8 for baseline production volume forecasts.

(5) Various feedstocks projected to be in use.

Again, however, demands for cellulosic biomass feedstocks (not now produced for any commercial purpose) to produce ethanol, or for soy biomass to manufacture bio-diesel could be extremely large, as Table 16 shows. These would translate into similarly large increases in use of agricultural land, as well as in water, fertilizer, and energy inputs. A variety of other oilseeds might replace soy as the current feedstock for the biomass component of bio-diesel, but some of these would require significant increases in current plantings and land area devoted to their cultivation. Increased fuel consumption for generating electricity would be proportional to the added demand for electrical energy required to replace gasoline use.

Because growing demands for electricity as a major energy source throughout the non-transportation sectors of the economy will already require substantial investments in added generating capacity, however, further increases in generating capacity and fuel use by electric utilities to meet the relatively modest added electric power demand for replacing projected gasoline would probably be manageable. Similarly, growing demand for CNG as a non-

Table 15. Alternative fuel volumes necessary to replace 25% of projected 2025 gasoline use.

Fuel	Form	Unit of Measure	Energy Density (Btu/unit) (1)	Fuel Energy (quadrillion Btu) (2)	Fuel Volume (billion units) (3)	% Increase from Baseline Production (4)
Ethanol (E90)	liquid	gallon	79,850	5.83	73.0	(5)
CNG	gaseous	cubic foot	928	5.50	5,921	14%
LPG	liquid	gallon	84,000	5.24	62.4	140%
Diesel	liquid	gallon	128,000	3.81	29.8	67%
Bio-Diesel (B20)	liquid	gallon	125,820	3.81	30.3	(5)
Electricity	--	kWh	11,765	4.50	382	8%
Hydrogen	gaseous	cubic foot	274	4.29	15,647	(6)

(1) Source: Wang, Michael, *GREET 1.5a -- Transportation Fuel-Cycle Model: Volume 1, Methodology, Use, and Results*, ANL/ESD-39, Argonne National Laboratories, 1999, Table 3.3 (low heating values).

(2) Calculated from forecast of light-duty vehicle travel reported in table 1 and energy efficiencies reported in Table 6.

(3) Equals total alternative fuel energy use divided by energy density of fuel.

(4) Equals total alternative fuel volume as a percent of production forecast for 2020 reported in U.S. Energy Information Administration, *Annual Energy Outlook 2000 -- Reference Case Forecast*, Appendix A, Tables A-1 and A-2. See Appendix Table A-9 for baseline production volume forecasts.

(5) Calculated percent increase is extremely large due to small projected baseline production.

(6) No production anticipated in absence of use as a transportation fuel.

transportation energy source means that the increased production and distribution capacity required for it to serve as a large-scale gasoline replacement (or as a feedstock for other alternative fuels) would represent a modest percentage increase in total natural gas production.

Fueling Infrastructure Requirements

A ubiquitous retailing or other distribution infrastructure – comparable to that now offered by the network of over 100,000 retail gasoline stations – would also be required to allow any alternative fuel to replace a significant fraction of gasoline use.[41] Refueling with a gasoline substitute would need to approach the convenience now afforded by gasoline retailing stations, including a wide choice of locations, speed and ease of refueling, convenient hours of operation, availability of ancillary services such as vehicle repair, and options for self service (possibly accompanied by driver education in safe fueling procedures). Refueling facilities would need to be available in a range of locations, including densely-developed urban areas, remote recreational locations, and at occasional points along highways in sparsely populated regions.

The existing distribution and retailing infrastructure for transportation fuels has been developed over nearly a century, with many facilities built in an era of readily-available land

Table 16. Feedstock demands for alternative fuel volumes necessary to replace 25% of projected 2025 gasoline use.

Fuel	Feedstock	Units of Measure		Fuel Volume (billion units) (1)	Feedstock Yield (fuel units per feedstock unit) (2)	Feedstock Demand (billion units) (3)	% Increase from Baseline Production (4)
		Fuel	Feedstock				
Ethanol (E90)	Cellulosic Biomass	gallon	ton	73.0	103	0.71	-- (5)
CNG	Natural Gas	cubic foot	cubic foot	5,921	0.959	6,177	15%
LPG	Petroleum	gallon	gallon	62.4	1.519	41.1	14%
LPG	Natural Gas	gallon	cubic foot	62.4	0.011	5,749	14%
Diesel	Petroleum	gallon	gallon	29.8	0.934	31.9	11%
Bio-Diesel (B20)	Petroleum	gallon	gallon	24.2	0.934	25.9	9%
	Soy Oil	gallon	pound	6.1	0.025	48.1	160%
Electricity	Projected U.S. Mix	kWh	various	382	-- (6)	-- (6)	-- (6)
Hydrogen	Natural Gas	cubic foot	cubic foot	15,647	2.025	7,726	18%

(1) Source: Table 15.

(2) Estimated from values reported in Wang, Michael, *GREET 1.5a -- Transportation Fuel-Cycle Model: Volume 1, Methodology, Use, and Results*, ANL/ESD-39, Argonne National Laboratories, 1999, Tables 3.3, 4.3, 4.11, 4.25, and 4.28.

(3) Equals alternative fuel volume divided by feedstock yield.

(4) Equals feedstock volume as a percent of production forecast for 2020 reported in U.S. Energy Information Administration, *Annual Energy Outlook 2000 -- Reference Case Forecast*, Appendix A, Tables A-1 and A-2. For biomass feedstocks, equals feedstock volume as a percent of total production for most recent year available, reported in U.S. Department of Commerce, Statistical Abstract of the United States, 1999, Tables 1129 and 1130. See Appendix Table A-9 for baseline production volume forecasts.

(5) Calculated percent increase is extremely large due to small projected baseline production.

(6) Various feedstocks projected to be in use.

and less concern for the potential environmental impacts of fuel handling and potential spillage. Investments in new or replacement infrastructure are likely to face far greater limitations on available land in and near developed areas, as well as detailed regulations on their design and operation to ensure user safety, health, and environmental protection. Siting of new fuel storage and retailing facilities is likely to be subject to more protracted public procedures than in the past, with the applicable regulations and permitting authority falling under the jurisdiction of a variety of federal, state, and local agencies. Safety, health, and environmental concerns – and the public's perception of how facility designs and operating procedures respond to these concerns -- can have major impacts on the acceptability of new facilities, the locations where they are allowed to be constructed, and their final costs.

Perhaps the most likely scenario for creating an expanded distribution network for certain alternative fuels would entail installation of separate storage capacity and pumping equipment for these fuels at some existing retail gasoline stations, as is now the case for diesel. For fuels such as ethanol and petroleum or bio-diesel, appropriate modifications might allow some existing gasoline storage tanks and pumps to be converted for use with these fuels, since gasoline storage and pumping demands would be reduced in proportion to the substitution of these fuels. In contrast, the physical properties of other alternative fuels such as natural gas or hydrogen can require specialized capabilities such as pressurized storage capacity and closed-coupling transfer systems to vehicles' fuel tanks, and providing these capabilities on a widespread basis is likely to require construction of at least some new, dedicated fueling facilities.

Potential Hazards from Alternative Fuels

Both conventional fuels such as gasoline and the various alternative fuels considered in this analysis can present potential safety, health, and environmental hazards. These include risks of fire or explosion in fuel processing and distribution, as well as potential hazards to human health from repeated or prolonged contact with fuels or their vapors. Potential environmental damages arising from fuel production, storage, or distribution, from normal vehicle operation and refueling, or from accidental fuel releases occurring at any point in the production, distribution, and fuel use cycle are another important concern.[42] Automobiles and trucks will continue to require high-density forms of on-board energy storage for the foreseeable future, which current vehicle technology limits to liquid and gaseous fuels or chemical batteries. These energy sources are each characterized by different potential risks in their production, storage, distribution, and use, and the specific properties that create these risks vary considerably among fuels.

Sources of Potential Hazards

Certain fuel-handling activities that are common to most fuels represent one major source of potential safety, health, and environmental hazards. For example, movement of fuels through transfer points in their distribution infrastructure -- pipeline junctions or loading facilities such as truck or barge terminals, for example -- provides locations where leaks can occur. Similarly, vehicle refueling at retail stations or other distribution points presents an obvious source of safety and environmental risks from fuel spillage, as well as of potential health risks to workers and vehicle owners from contact with fuel or recurring inhalation of fuel vapors.

For example, pressurized fuel distribution systems such as those required for CNG can pose not only environmental risks and fire hazards from leakage at junction and transfer points, but also risks to repair and emergency workers in cases of accidental mechanical failure of their components. As another illustration, leaks at critical points in a fuel's distribution infrastructure, at storage facilities, or from vehicles' on-board fuel tanks may allow volatile fuels such as gasoline or LPG to form potentially dangerous vapor clouds that can drift before dissipating or being ignited. The potential for vapors to reach dangerous concentrations is an especially important consideration within enclosed areas such as vehicle repair or parking facilities.

Physical or chemical properties of individual fuels represent other important sources of potential safety and health risks or hazards of environmental damage. Properties that directly affect the risks of fire in fuel handling and storage include the temperature at which a liquid fuels produces sufficient vapor concentrations to allow its ignition (called its flash point), a fuel's volatility (measured by the pressure exerted by fuel vapors in a closed container, an indication of its potential for leakage), its ignition temperature and flammability or explosion limits (the atmospheric concentrations at which a fuel will ignite or detonate), and the energy content of the fuel. The potential risks to human health from handling and use of fuels are determined by their toxicity from inhalation and skin contact, hazards from accidental ingestion, and risk of asphyxiation posed by oxygen displacement. A fuel's potential environmental impacts stem from toxic or criteria pollutant emissions that occur during its storage and distribution or during vehicle fueling and operation, as well as from damages to soil or

Table 17. Properties of gasoline and alternative fuels affecting potential safety and environmental hazards.

Fuel	Feedstock	Fuel Form	Boiling Point (°F at 1 atm.)	Vapor Pressure (psi)	Ignition Temperature (°F)	Explosion Limits (% gas in air)	Toxicity	Corrosiveness
Gasoline	Petroleum	liquid	200-220	7-16 at 100°F	428	1.4-7.6	Low, but effect of additives uncertain	Low
Ethanol	Biomass	liquid	172	2.3	793	4.3-19	Low, but affected by denaturant	Low
CNG	Natural Gas	gaseous	-259	2,400	1,004	5.3-15	Low, but some impurities may be toxic	Low, except for impurities
LPG	Petroleum or Natural Gas	liquid	-44	122 at 68°F	920-1,120	2.2-9.4	Low, but OSHA limits apply	Very low
Diesel	Petroleum	liquid	443-492	0.2 at 100°F	445	0.6-6.5	Low, but effect of additives uncertain	Low
Hydrogen	Natural Gas	gaseous	-4,320	--	1,050-1,080	4.1-74	Low	Low

Sources: U.S. Department of Energy, Energy Information Administration, *Alternatives to Traditional Transportation Fuels: An Overview*, DOE/EIA-0585/0, June 1994; U.S. Department of Transportation, *Summary Assessment of the Safety, Health, and Environmental and System Risks of Alternative Fuel*, DOT-VNTSC-FTA-95-5, Washington, D.C., March 1995; Battelle Memorial Institute, *Clean Air Program: Properties of Alternative Fuels*, UMTA-OH-06-0056-91-6, Columbus, OH, March 1996; U.S. Department of Transportation, *Infrastructure Implications of Next-Generation Vehicles* (Draft), March 1998.

water quality from its risk of spillage and toxicity.

Fuel properties such as flammability can pose significant risks to public safety, while other properties such as corrosiveness or fuel toxicity can damage the structural integrity of fuel handling or storage facilities, or present risks of contamination for environmental resources such as water supplies. Chemical or physical properties of fuels can also impose critical requirements for their safe handling, such as maintenance of pressurization throughout distribution systems for gaseous fuels, or careful temperature control during storage and distribution of highly volatile fuels.

Fuels requiring cryogenic cooling (such as natural gas or hydrogen in their liquefied states) can pose special problems from the effects of extremely low temperatures on materials used for storage and transportation facilities, and can also require elaborate systems to maintain those temperatures during fuel storage. The energy density of different fuels also indirectly affects the potential safety, health, and environmental risks posed by their widespread use, since fuels with lower energy content per unit volume require larger volumes of fuel to be produced and transported, thus increasing potential safety and environmental risks from transportation-related accidents.

Risk Characteristics of Specific Fuels

Table 17 summarizes important properties of gasoline and alternative light-duty vehicle fuels considered in this study that affect the potential safety, health, and environmental hazards they pose. Gasoline and alternative fuels each present differing risk profiles, as a consequence of differences in the properties that determine the seriousness of their potential risks to human health and safety and to the natural environment. These risk profiles clearly show that conventional fuels such as gasoline are not free from hazards, and that none of the alternative fuels considered in this study is inherently more "dangerous" than the fuels it would replace.

The existence of special safety concerns or particular hazards with alternative fuels that appear desirable because of the potential reductions in GHG emissions they offer (or their cost-effectiveness in achieving them) means that their safe and successful use will require intelligent planning of systems for their production and delivery. It will also require careful design and engineering of facilities for the production, storage, and distribution of alternative fuels, as well as of a network of outlets for refueling the large numbers of vehicles that would operate on any fuel that displaced a significant fraction of gasoline use. Finally, thorough training in the handling and use of alternative fuels by both workers involved in their production and distribution and vehicle owners using these fuels will be required to minimize potential hazards posed by their large-scale use.[43] Planning for the safe use of alternative fuels must emphasize using these efforts to eliminate obvious hazards in their routine production and delivery, as well as to minimize the probability and consequences of lower-probability incidents, since even their rare occurrence can have impacts on the use and acceptance of fuels extending well beyond the immediate consequences of these events.

Conclusions

The results of this analysis indicate that only a few alternative fuels appear to offer the potential to reduce GHG emissions from light-duty vehicle significantly from the levels generated by continued reliance on gasoline, regardless of the predominant vehicle and engine technology in use. Alternative fuels with the potential to do so include ethanol, diesel fuels derived entirely from petroleum or partly from biomass (bio-diesel), hydrogen, and possibly electricity.[44] Within the near-term time horizon of this study (approximately a decade), wider use of diesel, bio-diesel, corn-based ethanol, and electricity as light-duty vehicle fuels each appear to offer the potential for modest reductions in GHG emissions at realistic levels of gasoline replacement. Among these gasoline substitutes, however, only petroleum diesel appears likely to be cost-effective by comparison other strategies for reducing GHG emissions, and important concerns remain about whether light-duty diesel vehicles can comply with pending federal emission standards for certain criteria pollutants.

Over the longer term (a 20-25 year time horizon), ethanol derived from cellulosic biomass appears to offer the most realistic potential to achieve significant reductions in GHG emissions. This is partly because carbon emissions from its use as a fuel would be offset by carbon sequestration during growth of the biomass feedstock, but also partly because anticipated improvements in the energy efficiency of the process used to produce ethanol from cellulosic biomass. However, substituting ethanol for a significant fraction of gasoline use would require successful large-scale commercialization of currently experimental technology to produce it from herbaceous or woody biomass, as well as major investments in production, storage, and distribution facilities. As indicated previously, the most promising strategy for large-scale substitution of ethanol for gasoline appears to be widespread blending of ethanol with gasoline at a fraction below the commonly-assumed 85-90%, while continuing to utilize the current gasoline distribution and retailing infrastructure.

At the same time, advanced engine and drive system technologies – including direct injection internal combustion engines, hybrid internal combustion/electric drive systems, and fuel cells – could significantly reduce GHG emissions from their use over the longer term, by improving their energy efficiency and thus reducing total fuel energy consumption for light-duty vehicle travel. In combination with replacing a significant share of gasoline with certain alternative fuels, widespread use of these advanced technologies in light-duty vehicles offers the potential for very significant reductions in future GHG emissions from the level that would result from continued near-exclusive reliance on conventional internal combustion engines powered by gasoline.

Nevertheless, costs for replacing a significant fraction of gasoline use with any alternative fuel are likely to be very substantial. These include additional costs for manufacturing vehicles capable of operating on an alternative fuel, investments in new or expanded facilities to produce, store, and distribute it, and costs for obtaining the feedstock from which it is derived and for producing the fuel itself. The magnitude of these costs appears to make most alternative fuels unattractive strategies for reducing GHG emissions when their use is compared to measures to reduce GHG emissions from other sectors of the economy on the basis of cost-effectiveness.[45] The major exception appears to be petroleum-based diesel, which could modestly reduce GHG emissions and would require only minor modifications to existing gasoline production and distribution facilities. However, improvements in light-duty diesel engine designs and exhaust aftertreatment systems will be necessary in order to allow diesel vehicles to meet future emissions standards for criteria pollutants. Ethanol derived from cellulosic biomass appears to offer similar advantages, but its future production costs would need to be reduced significantly from currently projected levels in order for its widespread use to become a cost-effective strategy for reducing GHG emissions.

In short, only selected alternative transportation fuels appear to have the potential to reduce GHG emissions within the near term (approximately the next decade), and even these fuels would have to replace a major share of gasoline consumption by light-duty vehicles in order to achieve significant emissions reductions. Over the longer term (20-25 years), larger reductions in GHG emission are possible, but achieving them depends on the development and commercialization of new fuel production technologies, which remain somewhat uncertain. Because substituting most other fuels for gasoline fuels appears likely to be a costly strategy for reducing GHG emissions from motor vehicle use, the desirability of government policies promoting their use depends critically on the range of *other* strategies that are available to reduce GHG emissions -- including measures to reduce emissions from the non-transportation sectors of the U.S. economy – and on how the costs of these competing measures compare to those for producing and using alternative fuels.

Endnotes

[1] Automobiles and trucks powered by fuels other than gasoline or diesel account for only about 2% of all vehicles, so the assumption that virtually light-duty vehicles are fueled by gasoline appears to be justified; see Stacy C. Davis and Susan W. Diegel, *Transportation Energy Data Book: Edition 22*, Center for Transportation Analysis, Oak Ridge National Laboratory, ORNL-6967, September 2002 (http://www-cta.ornl.gov/cta/data/Index.html), Tables 6.3 and 9.1.

[2] Total GHG emissions include carbon dioxide (CO_2), methane (CH_4), and nitric oxide (N_2O), with the latter two weighted to their carbon dioxide equivalents using their estimated 100-year "Global Warming Potentials" (GWPs). The GWP of a greenhouse gas measures its potential contribution to global climate change over given future time period relative to that of an equivalent mass of CO_2; estimated 100-year GWPs for methane and nitric oxide are 21 and 310.

[3] This is an important distinction because under international GHG accounting conventions, emissions generated by vehicle operation are assigned to each nation's transportation sector, while those occurring during upstream activities are classified as emissions from industrial activity.

[4] The U.S. EPA argues that light-duty diesel vehicles should be able to meet the pending Tier 2 emission standards; see U.S. Environmental Protection Agency, Office of Mobile Sources, *Regulatory Impact Analysis – Control of Air Pollution from New Motor Vehicles: Tier 2 Motor Vehicle Emissions Standards and Gasoline Sulfur Requirements*, EPA420-R-99-023, December 1999 (http://www.epa.gov/otaq/tr2home.htm#ria), pp. IV-36 to IV-39. However, a recent study by the National Academy of Sciences notes that "the ability of this technology [direct-injection diesel engines] to comply with the upcoming Tier 2 and SULEV standards is highly uncertain." See National Academy of Sciences, *Effectiveness and Impact of Corporate Average Fuel Economy (CAFE) Standards*, ISBN 0-303-07599-8, Washington, D.C., National Academy Press, 2001, (http://books.nap.edu/books/0309076013/html), p. 3-12.

[5] Reliable estimates of these costs were available only for the near-term horizon, and required extensive efforts to convert them to a uniform per-vehicle basis denominated in current dollars.

[6] Additional vehicle production costs and investments in fuel production and distribution infrastructure are capital costs that must be amortized over appropriate lifetimes to determine their annualized equivalent values, while fuel production costs are recurring expenses associated with producing and using alternatives to gasoline.

[7] Stacy C. Davis and Susan W. Diegel, *Transportation Energy Data Book: Edition 22*, Center for Transportation Analysis, Oak Ridge National Laboratory, ORNL-6967, September 2002 (http://www-cta.ornl.gov/cta/data/Index.html), Tables 2.1 and 2.4.

[8] U.S. Environmental Protection Agency, *Inventory of GHG Emissions and Sinks: 1990-2000*, USEPA #236-R-02-003, April 2002, (http://yosemite.epa.gov/oar/globalwarming.nsf/content/ResourceCenterPublicationsGHGEmissionsUSEmissionsInventory2002.html), Tables 1-8 and 1-14.

[9] Computed from U.S. Environmental Protection Agency, Average Annual Emissions: All Criteria Pollutants (http://www.epa.gov/ttn/chief/trends/trends01/trends2001.pdf), Tables A-2, A-4, and A-5.

[10] Energy Policy Act of 1992, Public Law 102-486, October 24, 1992, Title III, Section 301.

[11] These forecasts are consistent with the "Reference Case" forecasts of vehicle use reported in the U.S. Energy Information Administration's *Annual Energy Outlook 2002*, December 2002, Table 7, http://www.eia.doe.gov/oiaf/aeo/results.

[12] As light trucks increasingly substitute for automobiles as passenger vehicles, their utilization rates (measured by annual miles driven per vehicle) -- which is now somewhat higher on average than automobiles -- is expected to converge toward that of automobiles. Thus the fractions of light-duty VMT accounted for by autos and light trucks during future years are assumed to be identical to the percentages of total vehicle registrations each represents.

[13] The Clean Air Act Amendments of 1990 required the use of reformulated gasoline (RFG) to reduce emissions of volatile organic compounds (VOCs) and nitrogen oxides (NOx) in areas that violate the federal air quality standard for ozone.

[14] For more information on EPA's Reformulated Gasoline and gasoline sulfur regulations, see http://www.epa.gov/otaq/rfg and http://www.epa.gov/otaq/tr2home.htm#documents.

[15] For details on EPA's diesel sulfur rule, see http://www.epa.gov/otaq/diesel.

[16] Vehicles capable of operating on methanol or ethanol as well as gasoline are usually referred to as flexible-fuel vehicles, and use the same fuel storage and delivery system regardless of the fuel in use. "Bi-fuel" vehicles can operate on compressed natural gas or gasoline, but have independent fuel storage and delivery systems for each of the two fuels. Various combinations of the proportion of bi-fuel vehicles in the light-duty fleet and the fraction of the time they actually operate on an alternative fuel could achieve the 10% gasoline replacement assumption, but at higher cost than "dedicated" natural gas vehicles.

[17] Depending upon how production costs for advanced-technology vehicles compare to those for internal-combustion powered vehicles, significantly higher fuel prices might be required to induce vehicle purchasers to pay the presumably higher prices for models employing advanced technology engines or drive systems.

[18] U.S. Environmental Protection Agency, *Inventory of GHG Emissions and Sinks: 1990-2000*, USEPA #236-R-02-003, April 2002, (http://yosemite.epa.gov/oar/globalwarming.nsf/content/ResourceCenterPublicationsGHGEmissionsUSEmissionsInventory2002.html), Tables 1-8 and 1-14.

[19] Carbon dioxide (CO_2), the main greenhouse gas produced by motor vehicle use, is a product of the combustion of the carbon compounds making up gasoline and other petroleum-based motor fuels.

[20] Extensive documentation of the model's development, calculation procedures, and data sources is provided in Argonne National Laboratories, *The Greenhouse Gas and Regulated Emissions from Transportation (GREET) Model*, Version 1.5, December 1999 (http://www.transportation.anl.gov/ttrdc/greet).

[21] Table 2 reports weighted average energy consumption rates for the mix of light-duty vehicles assumed to be in service during 2010, while those for the individual light-duty vehicle classes are given in Appendix Table A-1.

[22] The GREET model assumes that a blend of 80% petroleum-based diesel and 20% soy-based methyl ester – designated B20 -- is the form of bio-diesel most likely to be in commercial production and use.

[23] Table 3 shows composite GHG emission rates for the mix of light-duty vehicles in use during 2010; emission rates for sub-classes of light-duty vehicles are reported in Appendix Table A-2.

[24] This analysis focuses on criteria pollutant emissions from alternative fuels and gasoline during vehicle operation only (rather than over the full fuel cycle), because these are the subject of federal and California motor vehicle emissions regulations. The emissions rates for volatile organic compounds (VOC) produced by GREET include evaporative emissions from vehicles' fueling systems, which are regulated and tested separately, in addition to tailpipe emissions.

[25] Table 5 reports composite emission rates for the mix of vehicles assumed to comprise the light-duty fleet during 2010, while those for the individual sub-classes of light-duty vehicles are reported in Appendix Table A-3.

[26] With today's patterns of light-duty vehicle usage and fleet turnover, innovative engine technologies would need to be manufactured on a commercial scale approximately 10 years in advance of the study's longer-term horizon, or by the year 2015-20, in order to reach this level of use.

[27] Table 6 reports average energy use per vehicle-mile for the mix of automobiles, small light trucks, and large light trucks assumed to make up the U.S. light-duty vehicle fleet during 2025. Energy use per vehicle-mile for each of these three sub-classes of light-duty vehicles are shown in Appendix Table A-4.

[28] The figures reported in Table 7 are composite GHG emission rates for the mix of light-duty vehicles assumed to be in operation during 2025; Appendix Table A-5 shows separate GHG emissions rates for automobiles, small light trucks, and large light trucks.

[29] Criteria pollutant emission rates for individual sub-classes of light-duty vehicles operating on gasoline and alternative fuels are reported in Appendix Table A-6.

[30] Like the previous tables, all of the figures reported in Table 10 are calculated as weighted averages for the mix of automobiles, small light-duty trucks, and large light-duty trucks expected to comprise the light-duty vehicle fleet during the year 2025.

[31] The 25% usage figure for advanced engine technologies is an arbitrary target that is used for illustrative purposes, in the same way that the replacement of 25% of gasoline use by alternative fuels is used to illustrate their potential to reduce GHG emissions. No implication that costs to achieve 25% alternative fuel replacement of gasoline and 25% replacement of conventional ICEs with advanced–technology engines is intended, and in fact these costs are likely to differ substantially.

[32] Gasoline-powered fuel cell vehicles would employ on-board reforming of gasoline to produce hydrogen for use by a proton-exchange membrane fuel cell.

[33] Direct-injection internal combustion engines operating on diesel fuel would utilize compression ignition, a slightly different technology than gasoline-powered direct injection engines, which utilize spark ignition.

[34] This analysis assumes that the const-minimizing strategy will be for manufacturers to increase the on-board fuel storage capacity of vehicles designed to operate on alternative fuels to provide identical range to their gasoline-powered counterparts.

[35] For example, some infrastructure costs were originally expressed as costs per fueling station, for delivering some assumed number of gallons of fuel per month, or replacing gasoline distribution infrastructure for some percentage of light-duty vehicle travel. These were converted to a per-vehicle basis using the current average number of light-duty vehicles served per retail fueling station, average monthly mileage fuel use by light-duty vehicles, or the number of light-duty vehicles in use. Appendix Table A-7 reports the original estimates of these costs and their sources, and also provides details of how they were converted from their original forms to a per-vehicle basis and updated to reflect current prices.

[36] The annualized values of total capital investments associated with each alternative fuel were calculated assuming a 7% discount rate, and lifetimes of 15 years for vehicles and 25 years for fuel infrastructure.

[37] The fuel prices used to calculate gasoline and alternative fuel costs are year-2010 projections of delivered prices including taxes, reported by the U.S. Energy Information Administration (EIA) in its *Annual Energy Outlook 2002* "Reference Case" forecast. Taxes are subtracted from these figures because they do not represent real economic costs of fuel production. Because these forecasts assume increased future use of certain alternative fuels – principally natural gas and fuels derived using it as a feedstock – they may already incorporate some costs for expanded distribution infrastructure. Thus the total costs shown for these fuels in Table 16 may incorporate some "double-counting" of increased infrastructure costs.

[38] There are many possible ways that alcohol fuels could achieve the 20% gasoline displacement target other than the obvious strategy of having 20% of light-duty vehicles operating on these fuels at all times. For example, the target could also be met if 40% of the light-duty vehicle fleet were flexible-fuel vehicles capable of operating on these fuels but these vehicles were operated on gasoline half of the time. Part of the assumed gasoline displacement could also be accomplished at low cost by blending alcohol fuels with some or all gasoline at a low fraction, although this strategy is limited by both regulatory limits on total fuel oxygen content and potential operability problems with dedicated gasoline vehicles using alcohol fuel blends. We have not attempted to identify the least costly strategy for using alcohol fuels to achieve the gasoline replacement target used in this study.

[39] Forecast production volumes for each fuel and its feedstock during 2010 in the absence of efforts to promote its use as a substitute for gasoline are shown in Appendix Table A-8.

[40] The volume of each fuel and its feedstock projected to be produced during the long-term horizon year (2025) without efforts to promote its use as an alternative fuel are given in Appendix Table A-9.

[41] The U.S. Census Bureau reported that there were 126,889 retail gasoline stations in operation during 1997, of which 112,852 operated for the entire year; see U.S. Census Bureau, *1997 Economic Census of the United States: Retail Trade-Subject Series*, October 13, 2000, Table 1, p. 25.

[42] One such risk, that from emissions of criteria pollutants during fuel production and use, was previously discussed in Sections 4 and 5.

[43] For an extended discussion how these concerns can be addressed for specific fuels, see U.S. Department of Transportation and U.S. Department of Energy, *Summary Assessment of the Safety, Health, Environmental and System Risks of Alternative Fuels*, Volpe National Transportation Systems Center, Cambridge, Massachusetts, August 1995.

[44] Because it does not rely on carbon as a source of stored energy, hydrogen offers the potential to eliminate GHG emissions in vehicle operation (some emissions would still be generated in producing and transporting it), but its commercial-scale production and distribution within even the year-2025 long term time horizon used in this study remains highly uncertain.

[45] This is partly because many alternative strategies to reduce GHG emissions from non-transportation sources appear to be available at far lower costs, particularly if international trading of GHG-reduction credits as a source of emissions control is not arbitrarily limited.

Appendix

Table A-1. Energy consumption rates for LDV classes operating on near-term alternative fuel/engine technology combinations.

Fuel	Vehicle Energy Consumption (Btu/Mile)				Full Fuel Cycle Energy Consumption (Btu/Mile)			
	Auto	LDT1 (1)	LDT2 (2)	LDV Fleet (3)	Auto	LDT1 (1)	LDT2 (2)	LDV Fleet (3)
Conventional Gasoline	5,156	6,875	8,021	6,261	6,495	8,659	10,103	7,886
Reformulated Gasoline (Federal Phase 2)	5,156	6,875	8,021	6,261	6,549	8,732	10,187	7,952
Baseline (70%CG/30%RFG)	5,156	6,875	8,021	6,261	6,511	8,681	10,128	7,906
Ethanol (E85)	4,911	6,548	8,021	6,028	7,652	10,203	12,498	9,393
Methanol (M85)	4,911	6,548	8,021	6,028	7,461	9,948	12,187	9,159
CNG (Bi-Fuel Vehicle)	5,729	7,639	8,912	6,957	7,051	9,401	10,967	8,561
CNG (Dedicated Vehicle)	5,544	7,392	8,625	6,732	6,823	9,097	10,614	8,285
LNG	5,544	7,392	8,625	6,732	6,631	8,842	10,403	8,067
LPG	5,156	6,875	8,021	4,638	5,896	7,862	9,172	7,160
Petroleum Diesel	3,819	5,093	5,941	4,638	4,567	6,090	7,105	5,546
Bio-Diesel (B20)	3,819	5,093	5,941	4,638	4,728	6,305	7,358	5,742
Electricity	0	0	0	0	5,608	7,477	8,723	6,809

(1) Light-duty trucks with Gross Vehicle Weight Rating 6,000 pounds or less.

(2) Light-duty trucks with Gross Vehicle Weight Rating 6,001-8,500 pounds.

(3) Weighted average of Auto, LDT1, and LDT2 rates, computed using fleet shares shown in Table 1 as weights.

Source: Calculated using Wang, Michael Q., *GREET 1.5a -- Transportation Fuel-Cycle Model*, Argonne National Laboratories, January 2000.

Table A-2. Greenhouse gas emissions rates for LDV classes with near-term alternative fuel/engine technology combinations.

Fuel	Automobiles				Light-Duty Trucks 1 (1)				Light-Duty Trucks 2 (2)				Light-Duty Fleet (3)			
	CO_2 (4)	CH_4 (5)	N_2O (6)	GHGs (7)	CO_2	CH_4	N_2O	GHGs	CO_2	CH_4	N_2O	GHGs	CO_2	CH_4	N_2O	GHGs
Greenhouse Gas Emissions from Vehicle Operation Only (grams/vehicle-mile)																
Gasoline	390	0.084	0.028	401	521	0.09	0.033	533	607	0.09	0.04	622	474	0.087	0.032	486
Reform. Gasoline (Federal Phase 2)	361	0.077	0.028	372	482	0.083	0.033	494	562	0.083	0.040	576	439	0.080	0.032	450
Baseline (70% Gasoline/30%RFG)	382	0.082	0.028	392	509	0.088	0.033	521	594	0.088	0.040	608	463	0.085	0.032	475
Ethanol (E85)	83	0.126	0.028	94	111	0.135	0.033	124	139	0.135	0.040	154	103	0.131	0.032	115
Methanol (M85)	356	0.042	0.028	366	475	0.045	0.033	486	582	0.045	0.040	596	437	0.044	0.032	448
CNG (Bi-Fuel)	341	0.840	0.017	364	455	0.900	0.020	480	532	0.900	0.024	558	415	0.872	0.019	439
CNG (Dedicated)	330	0.840	0.022	355	441	0.900	0.026	468	514	0.900	0.032	543	401	0.872	0.025	427
LNG	326	0.840	0.022	350	435	0.900	0.026	462	508	0.900	0.032	536	396	0.872	0.025	422
LPG	369	0.109	0.028	380	492	0.117	0.033	505	574	0.117	0.040	589	448	0.113	0.032	460
Petroleum Diesel	307	0.011	0.016	312	410	0.014	0.024	417	478	0.017	0.032	488	373	0.013	0.022	380
Bio-Diesel (B20)	246	0.011	0.016	211	328	0.014	0.024	283	382	0.017	0.032	331	299	0.013	0.022	257
Electricity	0	0.000	0.000	0	0	0.000	0.000	0	0	0.000	0.000	0	0	0.000	0.000	0
Greenhouse Gas Emissions from Full Fuel Cycle (grams/vehicle-mile)																
Gasoline	475	0.665	0.0291	498	634	0.864	0.034	663	739	0.993	0.042	773	577	0.792	0.033	604
Reform. Gasoline (Federal Phase 2)	474	0.791	0.047	505	631	1.034	0.059	671	737	1.193	0.070	783	575	0.946	0.055	612
Baseline (70% Gasoline/30%RFG)	475	0.702	0.035	500	633	0.915	0.042	665	739	1.053	0.050	776	576	0.838	0.040	606
Ethanol (E85)	308	0.580	0.176	374	411	0.741	0.230	497	506	0.877	0.281	611	378	0.688	0.213	459
Methanol (M85)	464	0.675	0.029	487	619	0.890	0.035	648	758	1.080	0.042	794	570	0.821	0.034	597
CNG (Bi-Fuel)	428	2.085	0.018	478	572	2.560	0.022	632	667	2.837	0.026	735	521	2.384	0.021	577
CNG (Dedicated)	414	2.045	0.024	465	553	2.507	0.028	615	646	2.774	0.034	715	504	2.335	0.027	561
LNG	417	2.051	0.024	468	557	2.514	0.029	618	650	2.783	0.035	719	507	2.342	0.028	565
LPG	423	0.661	0.029	446	565	0.853	0.034	593	659	0.976	0.041	692	514	0.784	0.033	541
Petroleum Diesel	364	0.407	0.017	377	485	0.542	0.025	504	566	0.633	0.033	589	441	0.494	0.023	459
Bio-Diesel (B20)	309	0.328	0.019	282	412	0.436	0.028	377	481	0.510	0.036	441	375	0.398	0.025	343
Electricity	354	0.509	0.003	365	472	0.679	0.003	487	551	0.792	0.004	568	430	0.618	0.003	444

(1) Light-duty trucks with Gross Vehicle Weight Rating 6,000 pounds or less.
(2) Light-duty trucks with Gross Vehicle Weight Rating 6,001-8,500 pounds.
(3) Weighted average of Auto, LDT1, and LDT2 rates, computed using fleet shares shown in Table 1 as weights.
(4) Carbon dioxide; used as numeraire to express "Global Warming Potential" (GWP) of other GHGs.
(5) Methane; GWP relative to carbon dioxide = 21 (100-year value).
(6) Nitrous oxide; GWP relative to carbon dioxide = 310 (100-year value).
(7) CO_2 equivalent, with methane and nitrous oxide weighted by GWPs.

Source: Calculated using Wang, Michael Q., *GREET 1.5a – Transportation Fuel-Cycle Model* (with Volpe input assumptions), Argonne National Laboratories, January 2000.

Table A-3. *Criteria pollutant emission rates for LDV classes operating on near-term alternative fuel/engine technology combinations.*

Fuel	Automobiles					Light-Duty Trucks 1 (1)					Light-Duty Trucks 2 (2)					Light-Duty Vehicle Fleet (3)				
	VOC (4)	CO (5)	NOx (6)	SOx (7)	PM-10 (8)	VOC	CO	NOx	SOx	PM10	VOC	CO	NOx	SOx	PM10	VOC	CO	NOx	SOx	PM10
	Criteria Pollutant Emissions from Vehicle Operation Only (grams/vehicle-mile)																			
Gasoline	0.207	5.517	0.275	0.050	0.033	0.198	8.247	0.381	0.066	0.036	0.785	16.846	1.173	0.078	0.036	0.302	8.425	0.466	0.061	0.035
Reform. Gasoline (Federal Phase 2)	0.161	4.414	0.261	0.008	0.032	0.157	6.598	0.362	0.010	0.035	0.675	13.477	1.114	0.012	0.035	0.247	6.740	0.443	0.009	0.034
Baseline (70% Gasoline/30%aRFG)	0.193	5.186	0.271	0.037	0.033	0.186	7.752	0.375	0.050	0.036	0.752	15.835	1.155	0.058	0.036	0.286	7.919	0.459	0.045	0.034
Ethanol (E85)	0.176	3.310	0.248	0.010	0.026	0.168	4.948	0.343	0.013	0.027	0.589	7.581	0.997	0.016	0.027	0.243	4.625	0.409	0.012	0.026
Methanol (M85)	0.176	4.138	0.248	0.013	0.026	0.168	6.185	0.343	0.017	0.027	0.589	12.635	0.997	0.020	0.027	0.243	6.319	0.409	0.015	0.026
CNG (Bi-Fuel)	0.112	4.414	0.275	0.002	0.022	0.108	6.598	0.381	0.002	0.023	0.393	11.792	1.173	0.003	0.023	0.158	6.453	0.466	0.002	0.022
CNG (Dedicated)	0.045	4.414	0.248	0.002	0.022	0.047	6.598	0.343	0.002	0.022	0.204	10.108	1.173	0.003	0.022	0.073	6.167	0.439	0.002	0.022
LNG	0.045	4.414	0.248	0.002	0.022	0.047	6.598	0.343	0.002	0.022	0.204	10.108	1.173	0.003	0.022	0.073	6.167	0.439	0.002	0.022
LPG	0.077	4.138	0.248	0.000	0.022	0.084	6.185	0.343	0.000	0.023	0.456	12.635	0.997	0.000	0.023	0.144	6.319	0.409	0.000	0.022
Petroleum Diesel	0.080	1.070	0.600	0.048	0.121	0.091	1.139	0.600	0.064	0.121	0.540	1.208	1.224	0.075	0.130	0.162	1.118	0.706	0.058	0.123
Bio-Diesel (B20)	0.049	2.759	0.063	0.006	0.030	0.080	5.518	0.135	0.009	0.039	0.112	5.518	0.180	0.010	0.039	0.071	4.220	0.109	0.008	0.035
Electricity	0.000	0.000	0.000	0.000	0.021	0.000	0.000	0.000	0.000	0.021	0.000	0.000	0.000	0.000	0.021	0.000	0.000	0.000	0.000	0.021
	Criteria Pollutant Emissions from Full Fuel Cycle (grams/vehicle-mile)																			
Gasoline	0.296	5.646	0.461	0.164	0.051	0.316	8.420	0.630	0.219	0.060	0.923	17.047	1.463	0.256	0.064	0.410	8.582	0.692	0.200	0.056
Reform. Gasoline (Federal Phase 2)	0.278	4.580	0.531	0.164	0.091	0.313	6.819	0.722	0.218	0.114	0.857	13.735	1.534	0.255	0.127	0.389	6.942	0.770	0.199	0.105
Baseline (70% Gasoline/30%aRFG)	0.290	5.326	0.482	0.164	0.063	0.315	7.939	0.657	0.219	0.076	0.903	16.054	1.484	0.255	0.083	0.404	8.090	0.716	0.199	0.071
Ethanol (E85)	0.460	3.532	0.938	0.441	0.367	0.547	5.244	1.263	0.589	0.482	1.053	7.943	2.125	0.721	0.584	0.592	4.898	1.257	0.542	0.445
Methanol (M85)	0.238	4.364	0.462	0.068	0.037	0.252	6.486	0.629	0.090	0.042	0.691	13.004	1.348	0.110	0.045	0.320	6.596	0.673	0.083	0.040
CNG (Bi-Fuel)	0.142	4.576	0.623	0.118	0.034	0.148	6.814	0.845	0.157	0.038	0.440	12.045	1.714	0.183	0.041	0.195	6.651	0.888	0.143	0.037
CNG (Dedicated)	0.074	4.571	0.584	0.114	0.033	0.086	6.807	0.792	0.152	0.037	0.250	10.352	1.697	0.177	0.040	0.108	6.358	0.848	0.138	0.036
LNG	0.092	4.729	0.792	0.039	0.035	0.113	7.114	1.083	0.052	0.040	0.328	10.578	2.212	0.061	0.043	0.140	6.581	1.138	0.047	0.038
LPG	0.112	4.245	0.376	0.050	0.031	0.130	6.329	0.514	0.067	0.035	0.510	12.802	1.197	0.078	0.037	0.186	6.449	0.565	0.061	0.034
Petroleum Diesel	0.114	1.161	0.719	0.113	0.132	0.136	1.261	0.758	0.150	0.136	0.593	1.350	1.409	0.175	0.147	0.203	1.229	0.850	0.137	0.136
Bio-Diesel (B20)	0.127	2.901	0.250	0.058	0.044	0.184	5.707	0.384	0.077	0.061	0.234	5.739	0.471	0.090	0.061	0.166	4.392	0.336	0.070	0.052
Electricity	0.033	0.106	0.758	0.928	0.076	0.045	0.141	1.011	1.238	0.094	0.052	0.164	1.180	1.444	0.106	0.041	0.128	0.921	1.127	0.088

(1) Light-duty trucks with Gross Vehicle Weight Rating 6,000 pounds or less.
(2) Light-duty trucks with Gross Vehicle Weight Rating 6,001-8,500 pounds.
(3) Weighted average of Auto, LDT1, and LDT2 rates, computed using fleet shares shown in Table 1 as weights.
(4) Volatile organic compounds.
(5) Carbon monoxide.
(6) Nitrogen oxides.
(7) Sulfur oxides.
(8) Particulate matter 10 microns or less in diameter.

Source: Calculated using Wang, Michael Q., *GREET 1.5a – Transportation Fuel-Cycle Model* (with Volpe input assumptions), Argonne National Laboratories, January 2000.

Table A-4. Energy consumption rates for LDV classes operating on long-term alternative fuel/engine technology combinations.

Fuel	Engine Technology (1)	Vehicle Energy Consumption (Btu/Mile)				Full Fuel Cycle Energy Consumption (Btu/Mile)			
		Auto	LDT1 (2)	LDT2 (3)	LDV Fleet (4)	Auto	LDT1 (2)	LDT2 (3)	LDV Fleet (4)
Reformulated Gasoline	Spark-Ignition ICE	4,678	6,237	7,290	5,734	5,914	7,885	9,217	7,250
	Spark-Ignition Hybrid Electric	2,462	3,283	3,837	3,018	3,113	4,150	4,851	3,816
Ethanol (E90)	Spark-Ignition ICE	4,252	5,670	6,943	5,270	10,444	13,925	17,051	12,942
	Spark-Ignition Hybrid Electric	2,462	3,283	3,837	3,018	6,047	8,062	9,423	7,412
Methanol (M90)	Spark-Ignition ICE	4,252	5,670	6,943	5,270	6,602	8,803	10,779	8,181
	Spark-Ignition Hybrid Electric	2,462	3,283	3,837	3,018	3,822	5,096	5,957	4,686
CNG	Spark-Ignition ICE	4,455	5,940	7,290	5,523	5,382	7,176	8,807	6,673
	Spark-Ignition Hybrid Electric	2,752	3,669	4,288	3,373	3,324	4,432	5,181	4,075
LNG	Spark-Ignition ICE	4,455	5,940	7,290	5,523	5,542	7,389	9,069	6,871
	Spark-Ignition Hybrid Electric	2,752	3,669	4,288	3,373	3,423	4,564	5,335	4,196
LPG	Spark-Ignition ICE	4,252	5,670	6,943	5,270	4,855	6,473	7,926	6,016
	Spark-Ignition Hybrid Electric	2,752	3,669	4,288	3,373	3,141	4,189	4,896	3,851
Petroleum Diesel	Compression-Ignition ICE	3,118	4,158	4,860	3,823	3,813	5,084	5,942	4,674
	Compression-Ignition Hybrid Electric	2,034	2,712	3,170	2,493	2,487	3,316	3,875	3,048
Bio-Diesel (B20)	Compression-Ignition ICE	3,118	4,158	4,860	3,823	4,027	5,370	6,276	4,937
	Compression-Ignition Hybrid Electric	2,034	2,712	3,170	2,493	2,627	3,502	4,093	3,220
Electricity	Battery-Powered Electric Drive	0	0	0	0	3,587	4,783	6,389	4,541
Hydrogen (NG)	Proton-Exchange Membrane Fuel Cell	1,559	2,079	2,430	1,911	3,039	4,052	4,736	3,726
Hydrogen (Solar)	Proton-Exchange Membrane Fuel Cell	1,559	2,079	2,430	1,911	2,198	2,931	3,426	2,695

(1) ICE indicates internal combustion engine.

(2) Light-duty trucks with Gross Vehicle Weight Rating 6,000 pounds or less.

(3) Light-duty trucks with Gross Vehicle Weight Rating 6,001-8,500 pounds.

(4) Weighted average of Auto, LDT1, and LDT2 rates, computed using fleet shares shown in Table 1 as weights.

Source: Calculated using Wang, Michael Q., *GREET 1.5a -- Transportation Fuel-Cycle Model*, Argonne National Laboratories, January 2000.

Table A-5. Greenhouse gas emissions rates for LDV classes operating on long-term alternative fuel/engine technology combinations.

Fuel	Engine Technology (1)	Automobiles				Light-Duty Trucks 1 (2)				Light-Duty Trucks 2 (3)				Light-Duty Fleet (4)			
		CO₂ (5)	CH₄ (6)	N₂O (7)	GHGs (8)	CO₂	CH₄	N₂O	GHGs	CO₂	CH₄	N₂O	GHGs	CO₂	CH₄	N₂O	GHGs
		\multicolumn{16}{c}{Greenhouse Gas Emissions from Vehicle Operation Only (grams/vehicle-mile)}															
RFG	ICE	328	0.065	0.028	338	437	0.065	0.033	449	511	0.091	0.040	525	402	0.070	0.032	413
	Hybrid	172	0.065	0.028	182	230	0.065	0.033	242	269	0.091	0.040	283	211	0.070	0.032	223
Ethanol (E90)	ICE	49	0.098	0.028	60	65	0.098	0.033	77	82	0.137	0.040	97	61	0.104	0.032	73
	Hybrid	30	0.098	0.028	41	39	0.098	0.033	51	49	0.137	0.040	64	37	0.104	0.032	49
Methanol (M90)	ICE	310	0.033	0.028	319	413	0.033	0.033	424	506	0.046	0.040	519	384	0.035	0.032	395
	Hybrid	179	0.033	0.028	189	239	0.033	0.033	250	280	0.046	0.040	293	220	0.035	0.032	231
CNG	ICE	266	0.325	0.014	277	355	0.325	0.017	367	436	0.455	0.020	451	330	0.348	0.016	342
	Hybrid	164	0.325	0.014	175	219	0.325	0.017	231	256	0.455	0.020	272	201	0.348	0.016	214
LNG	ICE	263	0.325	0.014	274	350	0.325	0.017	362	430	0.455	0.020	446	326	0.348	0.016	338
	Hybrid	162	0.325	0.014	173	216	0.325	0.017	228	252	0.455	0.020	268	199	0.348	0.016	211
LPG	ICE	304	0.072	0.028	314	406	0.072	0.033	417	497	0.100	0.040	511	377	0.077	0.032	389
	Hybrid	197	0.072	0.028	207	262	0.072	0.033	274	307	0.100	0.040	321	241	0.077	0.032	253
Reformulated Diesel	ICE	252	0.011	0.016	257	336	0.014	0.024	343	392	0.017	0.032	403	309	0.013	0.022	316
	Hybrid	164	0.011	0.016	169	219	0.014	0.024	227	256	0.017	0.032	266	201	0.013	0.022	208
Bio-Diesel (B20)	ICE	206	0.011	0.016	211	275	0.014	0.024	283	321	0.017	0.032	331	253	0.013	0.022	260
	Hybrid	134	0.011	0.016	140	180	0.014	0.024	188	210	0.017	0.032	220	165	0.013	0.022	172
Electricity	Electric	0	0.000	0.000	0	0	0.000	0.000	0	0	0.000	0.000	0	0	0.000	0.000	0
Hydrogen (NG)	Fuel Cell	0	0.000	0.000	0	0	0.000	0.000	0	0	0.000	0.000	0	0	0.000	0.000	0
Hydrogen (Solar)	Fuel Cell	0	0.000	0.000	0	0	0.000	0.000	0	0	0.000	0.000	0	0	0.000	0.000	0
		\multicolumn{16}{c}{Greenhouse Gas Emissions from Full Fuel Cycle (grams/vehicle-mile)}															
RFG	ICE	426	0.711	0.043	454	568	0.926	0.054	604	664	1.097	0.064	706	522	0.861	0.051	556
	Hybrid	224	0.405	0.036	244	299	0.518	0.044	323	349	0.621	0.053	379	275	0.486	0.042	298
Ethanol (E90)	ICE	28	0.274	0.199	96	36	0.333	0.261	124	48	0.424	0.319	155	35	0.323	0.244	117
	Hybrid	18	0.200	0.127	62	23	0.234	0.165	79	30	0.296	0.194	96	22	0.230	0.153	74
Methanol (M90)	ICE	410	0.609	0.032	433	547	0.801	0.038	576	670	0.986	0.046	705	508	0.749	0.037	535
	Hybrid	237	0.366	0.030	254	317	0.477	0.036	338	370	0.565	0.043	395	291	0.444	0.035	311
CNG	ICE	329	1.283	0.015	361	439	1.602	0.018	478	539	2.022	0.022	588	408	1.536	0.017	446
	Hybrid	203	0.917	0.015	227	271	1.114	0.017	300	316	1.377	0.021	352	249	1.073	0.017	277
LNG	ICE	331	1.286	0.015	363	442	1.607	0.018	481	542	2.028	0.022	592	411	1.540	0.018	449
	Hybrid	204	0.919	0.015	228	273	1.117	0.018	302	318	1.380	0.021	354	251	1.076	0.017	278
LPG	ICE	349	0.527	0.029	369	466	0.679	0.034	490	570	0.844	0.041	601	433	0.641	0.033	456
	Hybrid	226	0.366	0.028	242	301	0.465	0.034	321	352	0.560	0.041	376	277	0.438	0.033	296
Reformulated Diesel	ICE	304	0.342	0.017	317	406	0.455	0.025	423	474	0.533	0.033	496	373	0.419	0.023	389
	Hybrid	198	0.227	0.016	208	264	0.302	0.025	278	309	0.353	0.033	327	243	0.278	0.022	256
Bio-Diesel (B20)	ICE	304	0.342	0.017	317	406	0.455	0.025	423	474	0.533	0.033	496	373	0.419	0.023	389
	Hybrid	176	0.217	0.018	186	235	0.289	0.026	249	274	0.339	0.035	292	216	0.266	0.024	229
Electricity	Electric	240	0.358	0.003	248	320	0.477	0.004	331	427	0.638	0.006	443	304	0.453	0.004	315
Hydrogen (NG)	Fuel Cell	183	0.417	0.001	192	244	0.556	0.002	256	285	0.650	0.002	299	224	0.512	0.002	235
Hydrogen (Solar)	Fuel Cell	41	0.089	0.001	44	55	0.119	0.001	58	65	0.139	0.001	68	51	0.109	0.001	53

(1) ICE indicates internal combustion engine.
(2) Light-duty trucks with Gross Vehicle Weight Rating 6,000 pounds or less.
(3) Light-duty trucks with Gross Vehicle Weight Rating 6,001-8,500 pounds.
(4) Weighted average of Auto, LDT1, and LDT2 rates, computed using fleet shares shown in Table 1 as weights.
(5) Carbon dioxide; used as numeraire to express "Global Warming Potential" (GWP) of other GHGs.
(6) Methane; GWP relative to carbon dioxide = 21 (100-year value).
(7) Nitrous oxide; GWP relative to carbon dioxide = 310 (100-year value).
(8) CO₂ equivalent, with methane and nitrous oxide weighted by GWPs.

Source: Calculated using Wang, Michael Q., *GREET 1.5a – Transportation Fuel-Cycle Model* (with Volpe input assumptions), Argonne National Laboratories, January 2000.

Table A-6. Criteria pollutant emission rates for LDV classes operating on long-term alternative fuel/engine technology combinations.

Fuel	Engine Technology (1)	Automobiles					Light-Duty Trucks 1 (2)					Light-Duty Trucks 2 (3)					Light-Duty Vehicle Fleet (4)				
		VOC (5)	CO (6)	NO$_x$ (7)	SO$_x$ (8)	PM-10 (9)	VOC	CO	NO$_x$	SO$_x$	PM10	VOC	CO	NO$_x$	SO$_x$	PM10	VOC	CO	NO$_x$	SO$_x$	PM10
		Criteria Pollutant Emissions from Vehicle Operation Only (grams/vehicle-mile)																			
Reform. Gasoline (Federal Phase 2)	ICE	0.125	2.759	0.036	0.007	0.031	0.125	2.759	0.036	0.009	0.031	0.158	5.518	0.135	0.011	0.041	0.131	3.253	0.054	0.009	0.033
	Hybrid	0.106	2.759	0.036	0.004	0.033	0.106	2.759	0.036	0.005	0.033	0.135	5.518	0.135	0.006	0.045	0.111	3.253	0.054	0.005	0.035
Ethanol (E90)	ICE	0.125	2.759	0.036	0.001	0.027	0.125	2.759	0.036	0.001	0.027	0.158	5.518	0.135	0.001	0.033	0.131	3.253	0.054	0.001	0.028
	Hybrid	0.106	2.759	0.036	0.001	0.030	0.106	2.759	0.036	0.001	0.030	0.135	5.518	0.135	0.001	0.038	0.111	3.253	0.054	0.001	0.031
Methanol (M90)	ICE	0.125	2.759	0.036	0.001	0.027	0.125	2.759	0.036	0.002	0.027	0.158	5.518	0.135	0.002	0.033	0.131	3.253	0.054	0.001	0.028
	Hybrid	0.106	2.759	0.036	0.001	0.030	0.106	2.759	0.036	0.001	0.030	0.135	5.518	0.135	0.001	0.038	0.111	3.253	0.054	0.001	0.031
CNG	ICE	0.059	1.655	0.036	0.001	0.023	0.059	1.655	0.036	0.002	0.023	0.072	4.414	0.135	0.002	0.025	0.061	2.150	0.054	0.002	0.023
	Hybrid	0.059	1.655	0.036	0.001	0.026	0.059	1.655	0.036	0.001	0.026	0.080	4.414	0.135	0.001	0.031	0.063	2.150	0.054	0.001	0.027
LNG	ICE	0.059	1.655	0.036	0.000	0.023	0.059	1.655	0.036	0.000	0.023	0.072	4.414	0.135	0.000	0.025	0.061	2.150	0.054	0.000	0.023
	Hybrid	0.059	1.655	0.036	0.000	0.026	0.059	1.655	0.036	0.000	0.026	0.080	4.414	0.135	0.000	0.031	0.063	2.150	0.054	0.000	0.027
LPG	ICE	0.065	1.655	0.036	0.000	0.023	0.065	1.655	0.036	0.000	0.023	0.088	4.414	0.135	0.000	0.025	0.069	2.150	0.054	0.000	0.023
	Hybrid	0.066	1.655	0.036	0.000	0.026	0.066	1.655	0.036	0.000	0.026	0.088	4.414	0.135	0.000	0.031	0.070	2.150	0.054	0.000	0.027
Petroleum Diesel	ICE	0.049	2.759	0.063	0.008	0.031	0.080	5.518	0.135	0.011	0.041	0.112	5.518	0.180	0.012	0.041	0.072	4.295	0.111	0.010	0.037
	Hybrid	0.049	2.759	0.063	0.005	0.031	0.080	5.518	0.135	0.007	0.041	0.112	5.518	0.180	0.008	0.041	0.072	4.295	0.111	0.006	0.037
Bio-Diesel (B20)	ICE	0.049	2.759	0.063	0.006	0.030	0.080	5.518	0.135	0.009	0.039	0.112	5.518	0.180	0.010	0.039	0.072	4.295	0.111	0.008	0.035
	Hybrid	0.049	2.759	0.063	0.004	0.030	0.080	5.518	0.135	0.006	0.039	0.112	5.518	0.180	0.007	0.039	0.072	4.295	0.111	0.005	0.035
Electricity	Electric	0.000	0.000	0.000	0.000	0.021	0.000	0.000	0.000	0.000	0.021	0.000	0.000	0.000	0.000	0.021	0.000	0.000	0.000	0.000	0.021
Hydrogen (NG)	Fuel Cell	0.000	0.000	0.000	0.000	0.021	0.000	0.000	0.000	0.000	0.021	0.000	0.000	0.000	0.000	0.021	0.000	0.000	0.000	0.000	0.021
Hydrogen (Solar)	Fuel Cell	0.000	0.000	0.000	0.000	0.021	0.000	0.000	0.000	0.000	0.021	0.000	0.000	0.000	0.000	0.021	0.000	0.000	0.000	0.000	0.021
		Criteria Pollutant Emissions from Full Fuel Cycle (grams/vehicle-mile)																			
Reform. Gasoline (Federal Phase 2)	ICE	0.229	2.908	0.251	0.106	0.082	0.264	2.957	0.323	0.141	0.099	0.321	5.749	0.470	0.165	0.121	0.259	3.435	0.317	0.130	0.095
	Hybrid	0.161	2.837	0.149	0.056	0.060	0.179	2.863	0.187	0.074	0.069	0.220	5.640	0.311	0.087	0.087	0.179	3.349	0.193	0.068	0.068
Ethanol (E90)	ICE	0.236	3.262	0.849	0.005	0.112	0.273	3.429	1.120	0.006	0.141	0.339	6.339	1.462	0.008	0.173	0.268	3.876	1.061	0.006	0.134
	Hybrid	0.170	3.050	0.507	0.003	0.079	0.192	3.147	0.664	0.004	0.095	0.234	5.971	0.869	0.004	0.115	0.190	3.610	0.631	0.003	0.092
Methanol (M90)	ICE	0.180	2.970	0.220	0.038	0.042	0.198	3.040	0.281	0.051	0.047	0.248	5.862	0.435	0.062	0.057	0.199	3.514	0.281	0.047	0.047
	Hybrid	0.138	2.881	0.142	0.022	0.038	0.149	2.922	0.178	0.029	0.041	0.184	5.708	0.301	0.034	0.051	0.150	3.403	0.184	0.027	0.042
CNG	ICE	0.075	1.775	0.271	0.058	0.031	0.080	1.815	0.349	0.077	0.034	0.098	4.611	0.519	0.094	0.039	0.081	2.298	0.345	0.071	0.034
	Hybrid	0.069	1.729	0.181	0.036	0.031	0.072	1.754	0.229	0.047	0.033	0.095	4.530	0.361	0.055	0.039	0.075	2.240	0.232	0.044	0.033
LNG	ICE	0.094	1.837	0.367	0.020	0.033	0.105	1.897	0.478	0.026	0.036	0.129	4.711	0.677	0.032	0.042	0.104	2.374	0.464	0.024	0.036
	Hybrid	0.080	1.767	0.241	0.012	0.032	0.088	1.805	0.309	0.016	0.034	0.113	4.589	0.454	0.019	0.041	0.089	2.287	0.305	0.015	0.035
LPG	ICE	0.093	1.744	0.135	0.032	0.030	0.102	1.773	0.167	0.043	0.033	0.133	4.559	0.296	0.052	0.037	0.104	2.259	0.176	0.040	0.033
	Hybrid	0.084	1.713	0.100	0.021	0.031	0.090	1.732	0.121	0.028	0.032	0.116	4.503	0.234	0.032	0.039	0.092	2.220	0.132	0.025	0.033
Petroleum Diesel	ICE	0.077	2.836	0.161	0.056	0.041	0.117	5.621	0.266	0.075	0.054	0.155	5.639	0.333	0.088	0.057	0.106	4.390	0.231	0.069	0.049
	Hybrid	0.067	2.810	0.127	0.037	0.038	0.104	5.585	0.220	0.049	0.051	0.140	5.597	0.280	0.057	0.051	0.094	4.357	0.190	0.045	0.045
Bio-Diesel (B20)	ICE	0.127	2.901	0.250	0.058	0.044	0.184	5.707	0.384	0.077	0.058	0.234	5.739	0.471	0.090	0.061	0.168	4.468	0.340	0.071	0.052
	Hybrid	0.100	2.851	0.185	0.038	0.039	0.148	5.641	0.298	0.050	0.051	0.191	5.662	0.370	0.059	0.053	0.134	4.408	0.261	0.046	0.046
Electricity	Electric	0.022	0.071	0.429	0.408	0.056	0.029	0.094	0.573	0.544	0.067	0.038	0.126	0.765	0.727	0.083	0.027	0.089	0.544	0.516	0.065
Hydrogen (NG)	Fuel Cell	0.015	0.128	0.208	0.057	0.029	0.020	0.171	0.278	0.076	0.032	0.023	0.200	0.325	0.088	0.034	0.018	0.157	0.255	0.070	0.031
Hydrogen (Solar)	Fuel Cell	0.007	0.035	0.126	0.061	0.027	0.010	0.047	0.167	0.082	0.029	0.011	0.055	0.196	0.095	0.031	0.009	0.043	0.154	0.075	0.029

(1) ICE indicates internal combustion engine.
(2) Light-duty trucks with Gross Vehicle Weight Rating 6,000 pounds or less.
(3) Light-duty trucks with Gross Vehicle Weight Rating 6,001-8,500 pounds
(4) Weighted average of Auto, LDT1, and LDT2 rates, computed using fleet shares shown in Table 1 as weights
(5) Volatile organic compounds.
(6) Carbon monoxide.
(7) Nitrogen oxides.
(8) Sulfur oxides.
(9) Particulate matter 10 microns or less in diameter

Source: Calculated using Wang, Michael Q., *GREET 1.5a – Transportation Fuel-Cycle Model* (with Volpe input assumptions), Argonne National Laboratories, January 2000.

Table A-7. Near-term light-duty vehicle manufacturing, fuel production, and fuel infrastructure costs for alternative fuel use.

Fuel	Feedstock	Added Vehicle Production Cost (1)			Fuel Infrastructure Costs (2)						Fuel Price ($/million Btu) (4)
		Original Estimate	in $ of	Equivalent in 2000 $ (3)	Original Estimate	in $ of	Equivalent in 2000 $ (3)	Basis for Estimate	Number of Vehicles	per-Vehicle Equivalent	
Gasoline	Petroleum	$0	2000	$0	$0	2000	$0	Baseline	--	$0	$7.64
Ethanol (E85)	Corn	$200	1990	$247	$59,181	1999	$60,442	33,000 gal/mo	438	$138	$15.37
Methanol (M85)	Natural Gas	$200	1990	$247	$59,181	1999	$60,442	33,000 gal/mo	325	$186	$9.22
CNG (Bi-Fuel)	Natural Gas	$964	1990	$1,192	$310,775	1989	$399,351	50,000 GGE/mo (one pump)	977	$409	$6.52
CNG (Dedicated)	Natural Gas	$526	1990	$650	$310,775	1989	$399,351	50,000 GGE/mo (one pump)	977	$409	$6.52
LNG	Natural Gas	$2,000	2000	$2,000	$0.193	1995	$0.211	$ per GGE	0.00013023	$1,616	$9.46
LPG	Petroleum/ Natural Gas	$198	1990	$245	$0.111	1995	$0.121	$ per GGE	0.00013280	$912	$10.94
Petroleum Diesel	Petroleum	$2,500	2000	$2,500	$25,000	2000	$25,000	33,000 gal/mo	977	$26	$5.95
Bio-Diesel (B20)	Petroleum/Soy	$2,500	2000	$2,500	$25,000	2000	$25,000	50,000 GGE/mo (one pump)	977	$26	$7.40
Electricity	Projected U.S. Mix	$17,000	2000	$17,000	$120	2000	$120	home charger (one/vehicle)	1	$120	$13.47

(1) Additional manufacturing costs for vehicles capable of operating on alternative fuel.
(2) Costs for additional fuel distribution, storage, and retailing infrastructure required to allow fuel to replace gasoline use.
(3) Adjusted to year-2000 dollars using change in Implicit Price Deflator for Gross Domestic Product.
(4) Source: Energy Information Administration, *Annual Energy Outlook 2002: Reference Case Forecast*, Table 3; converted to pre-tax value using fuel taxe rates reported in Federal Highway Administration, *Highway Statistics 1999*, Tables FE-21B and MF-121T.

Sources: Arthur D. Little, Inc., *Alternative Fuel Infrastructure Economics and Issues*, March 1997; Greene, David L., *An Assessment of Energy and Environmental Issues Related to the Use of Gas-to-Liquid Fuels in Transportation*, ORNL/TM-1999/258, Center for Transportation Analysis, Oak Ridge National Laboratory, November 1999; E.A. Mueller, Inc., *An Assessment of the Infrastructure Required to Refuel a Large Population of Natural Gas Vehicles*, January, 1989; U.S. Environmental Protection Agency, *Analysis of the Economic and Environmental Effects of Compressed Natural Gas as a Vehicle Fuel, Volume I: Passenger Cars and Light Trucks*, April 1990; Argonne National Laboratory, *Assessment of PNG Fuels Infrastructure: Phase 1 Report: Additional Capital Needs and Fuel-Cycle Energy and Emissions Impacts*, ANL/ESD/TM-140, January 1997, and *Assessment of PNG Fuels Infrastructure: Phase 2 Report: Additional Capital Needs and Fuel-Cycle Energy and Emissions Impacts*, ANL/ESD-37, August 1998; Williams, Larson, et al., *Methanol and Hydrogen from Biomass for Transportation, with Comparisons to Methanol and Hydrogen from Natural Gas*, PU/CEES Report No. 292, July 1995.

Table A-8. Near-term alternative fuel and feedstock production and energy content.

Fuel	Feedstock	Units of Measure		Baseline 2010 Production (quadrillion Btu) (1)		Btu Equivalent of Measurement Units (2)		Baseline 2010 Production (Billion Units) (3)	
		Fuel	Feedstock	Fuel	Feedstock	Fuel	Feedstock	Fuel	Feedstock
Gasoline	Petroleum	gallon	gallon	19.39	35.87	114,540	130,000	169	276
Ethanol (E85)	Corn	gallon	bushel	0.24	--	81,781	--	2.93	11.08
Methanol (M85)	Natural Gas	gallon	cubic foot	0.1	23.09	65,631	928	1.52	24,881
CNG	Natural Gas	cubic foot	cubic foot	0.23	23.09	928	928	248	24,881
LNG	Natural Gas	gallon	cubic foot	0.80	23.09	72,900	928	10.91	24,881
LPG	Petroleum	gallon	gallon	3.03	35.87	84,000	130,000	36	276
	Natural Gas	gallon	cubic foot	3.03	23.09	84,000	928	36	24,881
Petroleum Diesel	Petroleum	gallon	gallon	8.16	35.87	128,500	130,000	64	276
Bio-Diesel (B20)	Soy	gallon	pound	(4)	--	126,218	--	(4)	22.5
Electricity	Projected U.S. Mix	kwh	various	14.23	--	3,412	--	4,170	--

(1) Source: U.S. Energy Information Administration, Annual Energy Outlook 2002 -- Reference Case Forecast, Appendix A, Tables A-1 and A-2.

(2) Source: Wang, Michael, *GREET 1.5a -- Transportation Fuel-Cycle Model: Volume 1*, Methodology, Use, and Results, ANL/ESD-39, Argonne National Laboratories, 1999, Table 3.3 (low heating values).

(3) Computed from Baseline Production in quadrillion Btu divided by Btu equivalent of measurement units.

(4) No production anticipated in absence of use as a transportation fuel.

-- Indicates data not available.

Table A-9. Long-term alternative fuel and feedstock production and energy content.

Fuel	Feedstock	Units of Measure		Baseline 2025 Production (quadrillion Btu) (1)		Btu Equivalent of Measurement Units (2)		Baseline 2025 Production (Billion Units) (3)	
		Fuel	Feedstock	Fuel	Feedstock	Fuel	Feedstock	Fuel	Feedstock
Gasoline	Petroleum	gallon	gallon	23.21	37.59	114,540	130,000	203	289
Ethanol (E90)	Herbaceous Biomass	gallon	ton	0.17	(4)	81,781	--	2.08	(4)
Methanol (M90)	Natural Gas	gallon	cubic foot	0.29	24.09	65,631	928	4.46	25,958
CNG	Natural Gas	cubic foot	cubic foot	0.61	24.09	928	928	655	25,958
LNG	Natural Gas	cubic foot	cubic foot	0.91	24.09	72,900	928	12	25,958
LPG	Petroleum	gallon	gallon	3.45	37.59	84,000	130,000	41	289
LPG	Natural Gas	gallon	cubic foot	3.45	24.09	84,000	928	41	25,958
Petroleum Diesel	Petroleum	gallon	gallon	9.03	37.59	128,000	130,000	71	289
Bio-Diesel (B20)	Soy	gallon	pound	0.00	--	126,218	--	0	30.1
Electricity	Projected U.S. Mix	kWh	various	16.64	--	3,412	--	4,877	--
Hydrogen (liquid)	Natural Gas	gallon	cubic foot	0.00	24.09	30,100	928	0	25,958
Hydrogen (gaseous)	Natural Gas	cubic foot	cubic foot	0.00	24.09	274	928	0	25,958
Hydrogen (gaseous)	Water	cubic foot	cubic foot	0.00	--	274	--	0	--

(1) Source: U.S. Energy Information Administration, Annual Energy Outlook 2002 -- Reference Case Forecast, Appendix A, Tables A-1 and A-2.

(2) Source: Wang, Michael, *GREET 1.5a -- Transportation Fuel-Cycle Model: Volume 1*, Methodology, Use, and Results, ANL/ESD-39, Argonne National Laboratories, 1999, Table 3.3 (low heating values).

(3) Computed from Baseline Production in quadrillion Btu divided by Btu equivalent of measurement units.

(4) No production anticipated in absence of use as a transportation fuel or feedstock.

-- Indicates data not available.

www.ingramcontent.com/pod-product-compliance
Lightning Source LLC
Chambersburg PA
CBHW081845170526
45167CB00007B/2905